# Chapter 1: The Basics of Neural Oscillation

## What are Neural Oscillations?

Neural oscillations, often referred to as brainwaves, are the rhythmic or repetitive patterns of neural activity in the central nervous system. These oscillations occur as a result of the synchronized firing of neurons, which generate electrical impulses. Just like a symphony orchestra, the neurons in the brain work in harmony, producing waves of activity that oscillate at various frequencies. These neural oscillations are the foundation of how our brain processes information, regulates functions, and interacts with the world around us.

The study of these brainwave frequencies has led to a greater understanding of cognitive states and mental health. From states of deep sleep to high alertness, the frequencies of brainwaves help to define what we are experiencing at any given moment. The human brain, with its complexity, is capable of generating oscillations across a broad spectrum of frequencies, and each of these frequency bands has its unique role in mental and physical health.

## How Brainwaves Are Generated

The generation of brainwaves begins with neurons, the primary signaling cells in the brain. Neurons communicate with each other through electrical signals. When a group of neurons fire synchronously, their collective electrical activity produces oscillations that can be detected as brainwaves. These brainwaves can be measured using electroencephalography (EEG), which records the electrical activity on the scalp.

Brainwaves arise from the interplay between excitatory and inhibitory signals within the brain. When neurons are excited, they send electrical impulses that trigger other neurons, creating a ripple of electrical activity. If this ripple is synchronized across large groups of neurons, it results in an oscillation at a certain frequency. Different patterns of oscillation indicate different mental states, ranging from deep relaxation to intense concentration.

## The Importance of the Electrical Activity of the Brain

The brain is an electrically active organ, and its electrical activity is essential for all of its functions. Neurons communicate by transmitting electrical signals that allow the brain to process information, make decisions, and react to stimuli. The activity within a network of neurons is what enables the brain to generate complex thought processes and behaviors.

The brain's electrical activity is also responsible for regulating vital functions, such as heart rate, respiration, and digestion. Brainwaves influence our emotional state, cognitive functions, and even our overall health. For example, an imbalance in brainwave frequencies may contribute to conditions like anxiety, depression, or sleep disorders. Understanding and mastering neural oscillations can thus play a crucial role in maintaining optimal mental and physical health.

## Brainwave Patterns and Their Significance

Brainwaves are typically divided into five frequency bands, each associated with different cognitive states and behaviors. The most well-known of these are Alpha, Beta, Delta, and Theta waves. Each of these bands operates at different frequencies, and they are active in different mental states:

- **Alpha Waves (8–12 Hz):** These waves are typically associated with relaxation and calmness. They are most commonly observed when the brain is in a relaxed, yet alert state, such as during meditation or daydreaming. Alpha waves are often linked with creativity and deep relaxation.

- **Beta Waves (12–30 Hz):** Beta waves are linked with active thinking, concentration, and alertness. They dominate when a person is engaged in tasks that require focus and mental effort. Beta waves are typically present during problem-solving, decision-making, and other cognitive functions.

- **Delta Waves (0.5–4 Hz):** Delta waves are the slowest brainwaves, associated with deep sleep and healing. They are crucial for restorative functions, such as tissue repair, immune system strengthening, and memory consolidation. Delta waves are predominant during deep, dreamless sleep, allowing the body to repair and rejuvenate itself.

- **Theta Waves (4–8 Hz):** Theta waves are found during light sleep, deep meditation, and moments of deep creativity. These waves are associated with relaxation, introspection, and access to the subconscious mind. Theta waves are also linked to deep learning, intuition, and spiritual experiences.

Each brainwave frequency serves a distinct function and plays a critical role in our mental well-being. By understanding these brainwave patterns, we can learn to harness their power to enhance our cognitive abilities, regulate our emotions, and improve our overall health.

## The Goal of Mastering Brainwaves for Personal Growth and Well-being

Mastering brainwaves is about learning how to control and optimize the different frequency bands to enhance our physical, mental, and emotional well-being. Through targeted practices such as meditation, neurofeedback, and brainwave entrainment, we can learn to influence our brainwaves and shift our cognitive states.

For example, by enhancing Alpha waves, we can promote relaxation and creativity, making it easier to reduce stress and improve problem-solving. By regulating Beta waves, we can boost concentration and mental clarity, which is valuable in academic and professional settings. Furthermore, mastering Theta and Delta waves can enhance the depth of our meditation, improve sleep quality, and accelerate the body's natural healing processes.

Ultimately, mastering neural oscillations is a pathway to unlocking human potential. Whether through increased focus, creativity, or emotional balance, the ability to influence brainwaves provides an empowering tool for personal transformation. Understanding the science of brainwaves allows us to take control of our mental states, improve our physical health, and enhance our overall well-being.

As we dive deeper into each brainwave frequency and its practical applications, you will gain a clearer understanding of how neural oscillations shape your experience and how you can consciously master them for a more fulfilling life.

# Chapter 2: The Science Behind Brainwaves

## Neurophysiology and How Brainwaves are Measured

To truly understand how brainwaves function, it's essential to first explore the underlying neurophysiology. The brain is an organ of electrical activity, composed of billions of neurons, which are specialized cells responsible for transmitting electrical signals. These neurons communicate with each other through synapses, where electrical impulses travel across a small gap to activate neighboring cells. When large groups of neurons fire simultaneously, their collective electrical activity creates oscillations that can be measured and recorded.

The measurement of these electrical signals—brainwaves—requires a sensitive device capable of detecting minute electrical activity. This is where electroencephalography (EEG) comes in. EEG is a technique used to measure and record the electrical activity of the brain through sensors placed on the scalp. The data collected from EEG is displayed as a series of waveforms, which represent the brain's oscillations at different frequencies.

Brainwaves are categorized into distinct frequency bands, each corresponding to a specific state of consciousness or activity. These bands include the Alpha, Beta, Theta, Delta, and Gamma waves, each with its own frequency range. By studying the patterns and intensities of these waves, scientists and clinicians can gain insight into an individual's cognitive, emotional, and physiological states.

EEG provides a real-time window into the brain's electrical processes, allowing researchers and practitioners to investigate how different mental states, activities, or conditions impact the brain. It is one of the primary tools used in both neuroscience research and clinical practice for diagnosing conditions such as epilepsy, sleep disorders, and brain injuries.

## Electrophysiology and EEG (Electroencephalography)

The concept of electrophysiology, the study of the electrical properties of biological cells and tissues, is at the core of understanding brainwave generation. The electrical impulses generated by neurons result from changes in the ion concentration inside and outside the neuron. These ions—such as sodium ($Na+$), potassium ($K+$), calcium ($Ca2+$), and chloride ($Cl-$)—move in and out of the neuron through ion channels, creating a difference in electrical charge across the membrane.

When many neurons fire synchronously, the small electrical changes can build up into a detectable signal, creating a brainwave. The amplitude (height) and frequency (speed of oscillation) of the brainwave depend on various factors, including the number of neurons firing together and their synchronization. The EEG records this electrical activity, providing a graphical representation of the brain's oscillations over time.

EEG is commonly used to diagnose neurological conditions. For example, in patients with epilepsy, abnormal brainwave patterns are often seen as spikes or waves, which may indicate seizure activity. In sleep studies, EEG is used to monitor changes in brainwave patterns during different stages of sleep, such as REM (rapid eye movement) and non-REM sleep, which are vital for restorative functions.

## The Role of the Central Nervous System in Brainwave Generation

Brainwaves are primarily generated in the brain, but they are also influenced by the central nervous system (CNS), which includes the brain and spinal cord. The CNS coordinates all sensory and motor functions, and it plays a critical role in the regulation of brainwave activity.

The brain consists of several interconnected regions that contribute to the generation of brainwaves. The **thalamus**, for example, is a major relay station for sensory information, and it is integral in modulating brainwave frequencies. The **cortex**, the outermost layer of the brain, is where higher cognitive processes such as thinking, decision-making, and emotional regulation occur. The interaction between these regions, along with the rhythmic firing of neurons, creates the oscillatory brainwave patterns we observe on an EEG.

The autonomic nervous system (ANS), which controls involuntary functions like heart rate, digestion, and respiration, is also influenced by brainwaves. For example, certain brainwave states, such as those associated with relaxation (Alpha waves), can lower heart rate and blood pressure, promoting a state of calm. Conversely, heightened Beta wave activity can increase alertness and elevate stress responses, as seen during fight-or-flight situations.

The balance between the sympathetic and parasympathetic branches of the ANS, often referred to as the body's stress-response system, is linked to brainwave activity. Mastering brainwave states can therefore influence how the body reacts to stress, either facilitating relaxation or enhancing focus and alertness.

## Key Regions of the Brain Involved in Oscillations

While brainwaves are a result of the synchronized firing of neurons throughout the entire brain, certain areas are more prominently involved in specific oscillatory patterns. Understanding the regions responsible for different brainwave frequencies helps to explain the cognitive states they correlate with.

- **Frontal Lobe:** The frontal lobe is key in higher cognitive functions, such as decision-making, planning, problem-solving, and emotional regulation. Beta waves, which are associated with active thinking and focus, are often prominent in this region, especially during tasks that require concentration and mental effort.

- **Parietal Lobe:** This region is involved in sensory processing and spatial awareness. Theta waves, often linked with creativity, daydreaming, and access to the subconscious, are commonly generated here. The parietal lobe also plays a role in the transition between different stages of consciousness, such as from wakefulness to sleep.

- **Occipital Lobe:** The occipital lobe is primarily responsible for visual processing. Alpha waves, which are seen during relaxed but alert states, are commonly generated in this region when the brain is resting or in a meditative state.

- **Temporal Lobe:** The temporal lobe is associated with memory, auditory processing, and language comprehension. Delta waves, which dominate during deep sleep, are often produced here, particularly during the restorative stages of sleep that are critical for memory consolidation and learning.

- **Thalamus:** The thalamus, located deep within the brain, acts as a relay station, sending information between the brainstem and cortex. It is crucial in regulating the brain's overall rhythm and plays a central role in maintaining the synchronous firing of neurons that produce brainwaves.

The interconnected nature of these regions ensures that brainwave patterns are not confined to a single area, but rather reflect the collective activity of the brain. This synchronicity is vital for cognitive performance, emotional stability, and overall health. Disruptions in the synchronization of brainwaves can lead to cognitive impairments, mood disorders, and other neurological issues.

## Conclusion

In this chapter, we have delved into the neurophysiology of brainwaves, exploring the mechanisms behind their generation and how they can be measured using EEG. We've also examined the role of the central nervous system in regulating brainwave patterns, as well as the key brain regions involved in oscillations. Understanding the scientific foundations of brainwaves is essential for mastering them. In the next chapters, we will explore the individual brainwave frequencies—Alpha, Beta, Delta, and Theta—and discuss how each can be leveraged for personal growth, enhanced mental performance, and overall well-being.

By building on this scientific framework, we gain a deeper appreciation for the brain's remarkable ability to adapt and optimize itself. Mastering these frequencies allows us to harness the full potential of our brains, enabling us to achieve peak performance, enhance creativity, and improve our emotional and physical health.

# Chapter 3: Alpha Waves – Calm and Focus

## Defining Alpha Waves (8–12 Hz)

Alpha waves are one of the most well-known and studied brainwave frequencies, oscillating between 8 and 12 Hertz (Hz). They are typically associated with states of relaxation and calm, but also with a heightened sense of alertness and awareness. Alpha waves are most commonly observed when we are awake but relaxed, for example, when we are in a light meditative state, daydreaming, or even just resting with our eyes closed.

Despite the common misconception that the mind must be passive in order to experience Alpha waves, they are actually present when the brain is actively engaged in creative or relaxed thinking. For example, when solving a problem or brainstorming new ideas, Alpha waves can be abundant, allowing the brain to process information in a fluid, creative manner.

Alpha waves tend to dominate in moments when the brain is not overwhelmed by external stimuli or stress, providing a bridge between the active, alert states of Beta waves and the restful, restorative Delta waves. This makes Alpha waves particularly important for cognitive flexibility, creativity, and emotional regulation.

## The Role of Alpha Waves in Relaxation and Creativity

Alpha waves play a crucial role in promoting relaxation without making us feel drowsy. They facilitate the transition from a highly focused and analytical Beta state to a more relaxed but alert state. This frequency is associated with calmness, mental clarity, and a sense of mental rejuvenation.

In particular, Alpha waves are linked to states of creative flow. When the brain operates in an Alpha state, the "default mode network" (DMN)—a network of brain regions that is active when the mind is at rest or daydreaming—becomes engaged. This has been shown to foster insights, enhance problem-solving, and allow access to intuitive thinking. Many great discoveries and breakthroughs in science, art, and technology are associated with periods of relaxed focus, where the mind is calm but also able to creatively engage with new ideas.

Alpha waves are also thought to be helpful in managing stress. They reduce the impact of external stressors by fostering a calm yet alert mind. This allows individuals to deal with pressure in a way that is thoughtful and less reactive, a state that is conducive to clear decision-making and emotional balance.

## Practical Applications: Stress Reduction, Meditation, and Learning

Alpha waves are most commonly associated with practices that promote mental and emotional well-being, including:

- **Stress Reduction:** As mentioned, Alpha waves are linked with relaxation. One of the simplest ways to activate Alpha waves is through deep breathing exercises and progressive muscle relaxation techniques. These methods allow the body to enter a relaxed state, reducing the production of stress hormones like cortisol. When the brain enters an Alpha state, individuals are better able to manage stress and maintain their composure in tense situations.

- **Meditation:** Meditation is perhaps the most popular and effective way to activate Alpha waves. Many forms of meditation, including mindfulness meditation, guided imagery, and transcendental meditation, focus on inducing a relaxed, but focused state. These practices help individuals slow down their mental chatter, reduce stress, and access a deeper level of creativity and insight. People who regularly meditate often report improved emotional regulation, better decision-making, and enhanced focus.

- **Learning and Memory:** Alpha waves play a role in enhancing cognitive processing. Studies have shown that students who are in a relaxed state with an increased level of Alpha wave activity can retain and process information more effectively. The relaxed state promotes better attention and concentration, which in turn makes learning more efficient. Whether preparing for exams or mastering new skills, a relaxed, yet alert state of mind can significantly improve the capacity for learning.

Techniques for Increasing Alpha Wave Activity

While it may seem that Alpha waves naturally emerge when we are relaxed, there are a number of techniques and practices that can help increase the amount of Alpha activity in the brain. These methods not only help optimize mental well-being, but they also encourage personal growth and creativity.

- **Meditation and Mindfulness:** As mentioned, meditation is one of the most powerful tools for enhancing Alpha wave production. Regular meditation practice, whether focused on breathwork, body awareness, or mindfulness, has been shown to increase Alpha waves and improve overall brain function. Meditation sessions of 10 to 20 minutes can be a good starting point, gradually increasing the time spent in a meditative state.

- **Deep Breathing Exercises:** One of the most accessible and effective ways to activate Alpha waves is through deep, slow breathing. Techniques such as diaphragmatic breathing, box breathing (inhale for four counts, hold for four, exhale for four, and hold for four), and the 4-7-8 breathing technique (inhale for 4, hold for 7, exhale for 8) can quickly shift the brain into an Alpha-dominant state. These techniques help reduce sympathetic nervous system activity (the fight-or-flight response), allowing for relaxation and mental clarity.

- **Visualization and Guided Imagery:** Alpha waves are also strongly linked to visualization techniques, where one imagines peaceful, relaxing images or scenes, such as a beach or a forest. These images can help activate the Alpha state by quieting the mind and directing the focus inward. A guided imagery session—where a facilitator takes you through a series of calming visualizations—can significantly enhance Alpha wave production and deepen the relaxation process.

- **Listening to Binaural Beats or Isochronic Tones:** These are audio tracks designed to stimulate specific brainwave frequencies. Binaural beats work by presenting two different frequencies in each ear, creating the perception of a third frequency that matches the desired brainwave state. For Alpha waves, the difference between the two frequencies is typically between 8 Hz and 12 Hz. Isochronic tones, on the other hand, are single tones that pulse at specific intervals to entrain the brain to the desired frequency. Listening to Alpha brainwave music or sounds during relaxation, meditation, or even while working can enhance the production of Alpha waves.

- **Yoga and Physical Exercise:** Certain types of yoga, especially those that focus on deep stretching and breath control, can stimulate the parasympathetic nervous system, which increases Alpha wave activity. Practices like Yoga Nidra, or "yogic sleep," can induce deep relaxation and increase Alpha wave production. Similarly, physical exercise, particularly aerobic activities like walking, swimming, or cycling, can also encourage relaxation and trigger an increase in Alpha activity.

- **Nature and Environmental Factors:** Spending time in nature, especially in quiet, serene environments, can help reduce stress and promote Alpha waves. Natural surroundings have been shown to lower cortisol levels and enhance mental clarity. If spending time outdoors is not possible, simply listening to nature sounds—like birdsong, ocean waves, or rainfall—can have a similar effect.

Conclusion

Alpha waves are the brain's key to relaxation, creativity, and mental clarity. They help bridge the gap between a busy, analytical mind and a calm, reflective one. By increasing Alpha wave activity, individuals can tap into greater creativity, better stress management, and enhanced cognitive performance. As we've seen in this chapter, techniques like meditation, deep breathing, and guided imagery can significantly improve Alpha wave production, offering a path toward greater well-being and personal growth.

Mastering Alpha waves is not only about relaxation—it's also about optimizing brain function, accessing deeper creativity, and enhancing the overall quality of life. As we move forward in this book, we will explore how other brainwave frequencies, such as Beta, Delta, and Theta, each play a unique role in optimizing the brain for different aspects of life. But for now, fostering Alpha wave activity can provide a strong foundation for mastering neural oscillations and improving your mind-body connection.

# Chapter 4: Beta Waves – Focused Attention

### Defining Beta Waves (12–30 Hz)

Beta waves are the brain's most active frequency, ranging from 12 to 30 Hertz (Hz), and are most commonly associated with states of active thinking, concentration, and problem-solving. They represent the brain's engagement with the external environment, reflecting heightened alertness, focus, and cognitive effort. When Beta waves dominate the brain's activity, individuals are typically engaged in conscious, deliberate actions that require mental effort, such as work, studying, and decision-making.

While Beta waves are often seen as markers of the "busy mind," they are not inherently stressful. In fact, Beta activity is essential for cognitive performance, alertness, and intellectual engagement. However, prolonged or excessive Beta wave activity, especially in the higher end of the frequency range (above 20 Hz), can be associated with stress, anxiety, and overstimulation. Understanding how to optimize Beta waves can therefore improve cognitive performance without pushing the mind into a state of overload.

## The Role of Beta Waves in Concentration and Alertness

Beta waves play a critical role in focusing attention and managing mental tasks that require conscious thought and effort. Whether solving a math problem, having a conversation, or analyzing data, Beta waves facilitate the high-level cognitive processes necessary for goal-directed activity. These brainwaves help us stay alert and aware of our surroundings, allowing us to react quickly to stimuli and engage in complex tasks with clarity and focus.

In everyday life, Beta waves dominate during moments of active concentration. For example, when reading, driving, or solving a difficult puzzle, the brain is operating at its optimal frequency for deliberate thinking and information processing. Additionally, Beta waves are crucial when it comes to making decisions or weighing options, as they support logical reasoning and help manage working memory.

At the same time, the exact level of Beta activity needed depends on the task at hand. Moderate Beta activity supports optimal concentration and performance. Too little Beta wave activity, on the other hand, can make it difficult to focus or process complex information, while too much can lead to mental exhaustion or stress.

## Beta Waves and Problem-Solving

One of the most important functions of Beta waves is their role in problem-solving and complex cognitive processing. When confronted with a challenge, the brain engages Beta waves to facilitate the processing of multiple pieces of information simultaneously. This brainwave activity allows individuals to think logically, evaluate options, and make decisions with clarity.

For instance, Beta waves are dominant when you're solving puzzles, considering alternative solutions, or trying to understand complex systems. As Beta wave activity increases, the brain becomes more efficient at focusing on relevant information while ignoring distractions. This process is particularly important in analytical thinking, creative problem-solving, and the ability to adjust strategies based on changing circumstances.

In work environments where strategic thinking and problem-solving are critical, having optimized Beta wave activity can make a significant difference. People who can regulate and increase their Beta wave activity are often better equipped to stay focused on tasks and process large amounts of information efficiently. This cognitive advantage is particularly useful in fields like engineering, science, business, and law, where precision, analysis, and decision-making are fundamental.

## Managing Beta Activity for Optimal Performance

While Beta waves are essential for concentration and intellectual engagement, too much Beta activity can result in stress, anxiety, and mental fatigue. Prolonged overactivation of Beta waves, particularly in the high-frequency range (above 20 Hz), can lead to feelings of restlessness and an inability to relax. This is often seen in individuals who are under chronic stress or who are constantly multitasking. In such cases, the brain remains in a state of heightened alertness, and the body may even enter a fight-or-flight response, leading to negative health outcomes.

To maintain optimal Beta activity without succumbing to stress, it is important to find a balance. This can be done by incorporating relaxation practices and breaks throughout the day, ensuring that the brain does not become overtaxed. Engaging in mindfulness practices, taking deep breaths, and setting aside time for rest can all help reset the mind and prevent Beta wave activity from overwhelming the system.

Additionally, optimizing Beta wave production requires adopting strategies that enhance focus while preventing cognitive overload. Techniques such as time-blocking, task prioritization, and the Pomodoro technique (a time management method involving focused work intervals followed by short breaks) can help maintain a healthy level of Beta wave activity. These strategies allow individuals to stay productive without pushing the brain into overdrive.

## Using Beta Waves for Productivity

Beta waves are particularly helpful for increasing productivity, particularly in work environments or during tasks that require intense focus. By encouraging Beta wave activity, you can achieve enhanced cognitive performance and efficiency. However, understanding how to effectively use Beta waves involves knowing when to engage them and when to dial them back.

To increase productivity through Beta wave optimization, here are some key techniques:

- **Focused Work Sessions:** Use time management techniques such as the Pomodoro method to engage the brain in short bursts of focused activity. By working in focused intervals, you encourage Beta wave activity while allowing your brain to rest periodically to avoid mental fatigue.

- **Avoid Multitasking:** Multitasking often results in a fragmented focus, reducing overall productivity. Engaging in one task at a time allows Beta waves to function more effectively, providing sustained attention to the task at hand.

- **Mindfulness for Focus:** Mindfulness practices—such as mindful breathing, body scans, or sensory awareness—can help anchor attention and maintain focused Beta wave activity. These practices allow individuals to stay present and engaged without becoming overwhelmed by extraneous thoughts or distractions.

- **Cognitive Training Games:** Certain cognitive exercises and brain training games are designed to enhance Beta wave activity by challenging the brain with problem-solving tasks, memory games, and puzzles. Regular engagement with such exercises can sharpen focus and improve working memory, further enhancing productivity.

- **Brainwave Entrainment:** Using brainwave entrainment methods such as binaural beats or isochronic tones can also help tune the brain to the optimal Beta frequency range. Listening to Beta-specific audio tracks can help boost focus, mental clarity, and performance, especially when preparing for tasks requiring high levels of concentration.

## Conclusion

Beta waves are essential for activities that require mental engagement, from concentration and problem-solving to creative thinking and decision-making. While they are necessary for focused work and productivity, it is equally important to manage Beta activity in a way that prevents burnout or overstimulation. By optimizing Beta waves, you can enhance cognitive performance, improve concentration, and boost productivity without allowing stress and anxiety to take over.

In this chapter, we have explored the role of Beta waves in promoting focused attention, logical thinking, and problem-solving. We've also discussed strategies for optimizing Beta wave activity, such as mindfulness practices, cognitive training, and time management techniques. By understanding and mastering Beta waves, you can unlock your brain's full potential, creating a productive and balanced approach to work, problem-solving, and daily tasks.

As we continue to explore other brainwave frequencies, we'll examine how Delta and Theta waves contribute to deep rest, creativity, and healing—offering a well-rounded approach to mastering neural oscillations for both personal and professional growth.

# Chapter 5: Delta Waves – Deep Rest and Healing

Defining Delta Waves (0.5–4 Hz)

Delta waves, the slowest of all brainwave frequencies, range from 0.5 to 4 Hertz (Hz). They are primarily associated with deep, restorative sleep and are most prominent during stages of deep, non-REM (Rapid Eye Movement) sleep. While Delta waves are linked to sleep, they also play a significant role in healing, regeneration, and memory consolidation. These brainwaves are crucial for both physical and mental restoration, allowing the body to repair itself and the mind to process and store new information.

Unlike Alpha or Beta waves, which are present when the brain is active and engaged, Delta waves signify the brain's transition into a state of rest and deep healing. During deep sleep, the body enters a parasympathetic state—the "rest and digest" mode—where heart rate slows, muscles relax, and healing processes are accelerated. The presence of Delta waves is the brain's signal that it is entering a state of deep rest, which is essential for overall health and longevity.

## The Importance of Delta Waves in Deep Sleep and Physical Healing

Delta waves are integral to the most restorative phase of sleep: deep sleep. This stage of sleep is often referred to as slow-wave sleep (SWS), during which the brain exhibits large, slow Delta waves. Deep sleep is crucial for a variety of physiological functions, including:

- **Cellular Repair and Regeneration:** Delta waves are directly linked to tissue repair, immune function, and cellular regeneration. The body releases growth hormones during this stage, aiding in the repair of damaged tissues, building new muscle tissue, and replenishing cells throughout the body. This is especially important for athletes, people recovering from illness or injury, and anyone aiming to maintain physical vitality.

- **Memory Consolidation and Learning:** Deep sleep is also essential for memory consolidation. During the deep sleep stages dominated by Delta waves, the brain processes and solidifies information learned throughout the day. This process is vital for long-term memory retention and helps convert short-term memories into stable, long-term ones. For students, professionals, and anyone trying to enhance learning capabilities, prioritizing deep sleep is key to improving cognitive function.

- **Detoxification of the Brain:** The brain uses deep sleep to clear waste products that accumulate throughout the day, including beta-amyloid, a protein that is often linked to Alzheimer's disease. The deep sleep induced by Delta waves activates the glymphatic system, which helps eliminate these waste materials, allowing the brain to "cleanse" itself overnight.

In short, Delta waves are essential for physical restoration, cognitive maintenance, and overall health. These waves facilitate critical processes that occur during deep sleep, providing the body with the time and space it needs to heal, regenerate, and consolidate memories.

## How Delta Waves Are Linked to Restorative Processes in the Body

The restorative functions of Delta waves go beyond sleep; they also affect the body's physical and mental health. Delta waves promote an optimal environment for healing and repair by:

- **Boosting Immune Function:** The deep sleep associated with Delta waves helps strengthen the immune system by stimulating the production of cytokines, proteins that help fight off infection and inflammation. During Delta wave activity, the body is more efficient at combating illness and repairing damaged tissues, aiding in the body's natural healing process.

- **Hormonal Regulation:** Delta waves influence the production of growth hormone, which is critical for tissue repair, muscle building, and fat metabolism. The growth hormone released during deep sleep is also linked to the regeneration of the brain's cells, helping improve brain function over time. Additionally, Delta waves help regulate other hormones, contributing to a balanced metabolic function.

- **Reducing Stress:** Since Delta waves dominate during the deepest stages of sleep, they allow the body and mind to reset and recover from the stressors encountered throughout the day. Stress is known to impair the immune system and disrupt sleep patterns, but Delta wave activity helps the body unwind, promoting emotional and physical balance. By encouraging Delta wave activity, the body can reach a true state of relaxation and rejuvenation.

- **Balancing the Nervous System:** Delta waves are a signal of parasympathetic nervous system activity, which helps the body shift from the "fight or flight" sympathetic state to a more restful and healing state. By promoting Delta waves, the body achieves a state of homeostasis, which is vital for preventing chronic conditions linked to long-term stress, such as cardiovascular disease and high blood pressure.

## Leveraging Delta Waves for Sleep Optimization and Healing

Since Delta waves are most active during deep sleep, they play a key role in promoting restorative rest. However, many people struggle with achieving the amount of deep sleep necessary for optimal health. Poor sleep quality, stress, and various health conditions can prevent the brain from entering deep, Delta-dominant sleep stages. Fortunately, there are strategies to optimize Delta wave activity and improve the quality of sleep:

- **Prioritize Sleep Hygiene:** The first step in optimizing Delta waves is to improve sleep hygiene. This includes maintaining a regular sleep schedule, creating a dark and quiet sleep environment, and eliminating distractions like screens before bedtime. A consistent sleep routine ensures that the body and brain have enough time to enter the deep sleep stages where Delta waves thrive.

- **Relaxation Techniques:** Before bed, practicing relaxation techniques can help promote the transition into Delta sleep. Techniques such as progressive muscle relaxation, deep breathing, and mindfulness meditation can help calm the mind and prepare the body for deep rest. These practices reduce the production of cortisol, a stress hormone, and encourage the onset of Delta wave activity.

- **Mindfulness and Meditation:** Mindfulness practices, especially guided meditations aimed at inducing deep relaxation, can increase Delta wave activity. Meditation encourages the brain to quiet its thoughts, facilitating the onset of deep sleep and ensuring that Delta waves are more prominent during the night.

- **Brainwave Entrainment:** Listening to brainwave entrainment audio, such as binaural beats or isochronic tones designed to stimulate Delta waves, can help guide the brain into deep sleep. These audio tracks use sound frequencies that match the brain's natural Delta wave rhythms, helping the brain synchronize with these frequencies and promoting a more restful, deep-sleep experience.

- **Exercise:** Regular physical exercise is one of the most effective ways to enhance Delta wave activity during sleep. Exercise helps reduce anxiety, stress, and physical tension, which are often barriers to deep sleep. It also promotes the release of growth hormones, further enhancing the body's natural healing processes. However, it's important to avoid intense exercise close to bedtime, as this can have the opposite effect, raising energy levels and disrupting sleep.

- **Diet and Nutrition:** Certain foods and supplements can also support Delta wave production. For example, magnesium, found in foods such as leafy greens, nuts, and seeds, has been shown to help the body relax and promote deeper sleep. Melatonin-rich foods like cherries or melatonin supplements can also encourage the onset of sleep and enhance Delta wave activity.

## Delta Waves and Healing Beyond Sleep

While Delta waves are most prominent during sleep, their influence extends beyond the realm of rest. Delta waves are involved in numerous restorative processes that enhance overall well-being. These include the healing of physical injuries, mental rejuvenation, and the reduction of stress.

- **Physical Recovery:** Athletes, for instance, rely on Delta waves to help repair muscle tissue and recover from the physical demands of training. The body produces the most significant amount of growth hormone during Delta sleep, which aids in tissue repair and the rebuilding of muscles and bones.

- **Mental Rejuvenation:** Delta waves also facilitate mental healing. People who experience emotional trauma, high levels of stress, or burnout can benefit from increased Delta wave activity during sleep. This restoration enables the mind to process and integrate emotional experiences, allowing for psychological healing and emotional balance.

- **Accelerating the Healing Process:** By leveraging techniques that increase Delta wave activity, individuals can enhance the body's ability to recover from illness or surgery. The deep restorative sleep facilitated by Delta waves provides the optimal environment for immune function, tissue regeneration, and overall health recovery.

## Conclusion

Delta waves are crucial for deep rest, healing, and restoration. Dominating during the most profound stages of sleep, they support essential functions such as cellular repair, memory consolidation, immune function, and emotional well-being. Understanding how to enhance Delta wave activity—through improving sleep quality, using relaxation techniques, and incorporating lifestyle changes—can have a significant impact on overall health and recovery.

By prioritizing Delta waves, we not only improve the quality of our sleep but also foster a state of deep physical and mental rejuvenation that supports long-term health and vitality. As we continue to explore other brainwave frequencies, the role of Theta waves in creativity and deep meditation will further demonstrate the complex and interconnected nature of our brain's oscillations, creating a comprehensive approach to brainwave mastery.

# Chapter 6: Theta Waves – Creativity and Deep Meditation

### Defining Theta Waves (4–8 Hz)

Theta waves are brainwave frequencies that fall within the range of 4 to 8 Hertz (Hz), and they are often associated with deep states of relaxation, creativity, and introspection. Theta waves emerge when the brain is transitioning between light sleep and deep sleep, as well as during deep meditation or intense states of creative flow. These waves are essential for accessing deep levels of the subconscious mind, and they represent a bridge between conscious thought and unconscious processes.

Theta waves are typically present during the lighter stages of sleep but can also be observed during the awake state when a person is deeply relaxed or meditative. They occur when the brain begins to slow down from the higher frequencies associated with waking states (such as Beta and Alpha waves), allowing the individual to enter a profound state of mental stillness. This state is not only ideal for relaxation but also for unlocking creative potential and deep emotional processing.

## The Role of Theta Waves in Deep Meditation and Creativity

Theta waves are closely linked with deep meditation, a state where individuals are able to tap into their subconscious mind and experience profound clarity. The calming, slow rhythm of Theta waves allows the mind to access deeper levels of thought, intuition, and insight. When in a Theta-dominant state, the mind becomes more receptive to creative ideas, innovative solutions, and spontaneous insights.

- **Deep Meditation:** In meditation, Theta waves promote a state of calm that allows the practitioner to transcend ordinary thought patterns and access a deeper level of awareness. Meditation techniques such as mindfulness, guided visualization, and transcendental meditation can induce Theta wave activity, helping individuals enter a space of pure presence. In this state, mental chatter fades away, and the mind becomes still—this is often when individuals experience feelings of deep connection, peace, or spiritual insight.

- **Creativity and Problem-Solving:** Theta waves are known to be especially helpful in fostering creativity. When the brain operates in a Theta-dominant state, it can more easily break away from conventional thinking and begin to form new, innovative ideas. Many artists, musicians, writers, and entrepreneurs have credited Theta waves for their creative breakthroughs. This state allows individuals to access inspiration from their subconscious, making it easier to approach problems from fresh perspectives and unlock new solutions.

- **Intuition and Inner Wisdom:** Theta waves also allow for greater access to intuitive knowing. In Theta states, the brain becomes more attuned to subtle cues and inner wisdom, making it easier to trust gut feelings or instincts. This is why many people find that their most profound ideas and insights often come during moments of stillness or relaxation, when the mind is calm enough to listen to the subconscious.

- **Emotional Healing and Integration:** Theta waves play an important role in emotional processing and healing. During deep meditative states, Theta waves allow individuals to access deeply buried emotions and unresolved issues, facilitating emotional release and integration. This makes Theta wave activity particularly useful for individuals seeking emotional clarity, inner peace, or personal transformation.

## Theta Waves and Access to the Subconscious Mind

One of the most powerful aspects of Theta waves is their ability to connect the conscious mind with the subconscious. The subconscious mind holds much of our stored memories, emotions, and automatic thought patterns. When Theta waves dominate, it becomes easier to access these deeply stored parts of the mind and bring unconscious thoughts and feelings to the surface.

- **Memory and Learning:** Theta waves are linked to enhanced memory and learning ability. During periods of Theta activity, the brain can process and integrate information more deeply. This is particularly relevant for individuals who are trying to learn new skills, absorb complex information, or heal from past emotional trauma. By entering a Theta state, one can better consolidate and retain new memories, enhancing overall cognitive function.

- **Access to Deep Thoughts and Insights:** The Theta state also promotes introspection and self-awareness. By quieting the mind, Theta waves allow individuals to reflect on their life and make sense of their experiences. Many people use Theta meditation to explore their beliefs, uncover their life's purpose, or gain insight into their relationships. This reflective state provides a gateway to greater self-understanding and clarity.

- **Therapeutic and Healing Applications:** Theta waves are frequently used in therapeutic settings to assist with overcoming emotional challenges, anxiety, and trauma. Techniques like Theta Healing™ have been developed to access these deep subconscious states, facilitating transformative emotional healing. By engaging with the subconscious mind in a relaxed, Theta-dominant state, individuals can reprogram limiting beliefs, release emotional blockages, and achieve greater mental clarity and peace.

## Techniques to Increase Theta Activity for Personal Growth

While Theta waves are naturally occurring during deep meditation and sleep, there are specific practices that can encourage Theta wave production during waking hours. By incorporating these techniques into daily life, individuals can unlock their creative potential, enhance emotional healing, and access a deeper state of mindfulness.

- **Meditation Practices:** Meditation is one of the most effective ways to increase Theta wave activity. Practices such as mindfulness meditation, progressive relaxation, and deep breathing help quiet the mind and slow brainwave activity, allowing Theta waves to emerge. In particular, practices such as **Transcendental Meditation (TM)** and **Yoga Nidra** are known to induce deep Theta states. These forms of meditation guide the practitioner into a relaxed state where the conscious mind steps aside, allowing deeper subconscious insights to surface.

- **Visualization and Guided Imagery:** Visualization techniques are another powerful way to increase Theta wave activity. By vividly imagining calming or inspiring images—such as scenes of nature, healing light, or peaceful landscapes—the brain can enter a Theta-dominant state. Guided imagery sessions, where a facilitator leads you through a series of mental images, can also promote Theta waves and deepen the meditative experience.

- **Binaural Beats and Isochronic Tones:** Binaural beats and isochronic tones are auditory tools designed to entrain the brain to specific frequencies. Listening to music or soundscapes that are tuned to Theta frequencies (typically between 4 and 8 Hz) can help the brain synchronize with these wavelengths. Binaural beats, for instance, work by presenting two slightly different frequencies in each ear, which the brain then perceives as a single, unified sound. These auditory cues can guide the brain into a Theta state, enhancing creativity, relaxation, and deep meditation.

- **Deep Breathing and Relaxation Exercises:** Relaxation techniques, such as deep diaphragmatic breathing or body scans, promote Theta wave production by slowing the body's stress response and calming the mind. Deep breathing, especially at a slow and steady rhythm, can help lower brainwave frequency and induce a Theta state. The 4-7-8 breathing technique, where you inhale for four counts, hold for seven, and exhale for eight, is particularly effective for transitioning into Theta waves.

- **Hypnosis and Self-Hypnosis:** Hypnosis is a process of guiding the mind into a deeply relaxed, Theta-dominant state. By using relaxation and suggestion techniques, a skilled practitioner can help individuals access the subconscious mind for therapeutic purposes. Self-hypnosis techniques also allow individuals to enter this Theta state independently, facilitating personal growth, emotional healing, and creativity.

## Theta Waves and Personal Growth

Theta waves offer significant potential for personal growth, both emotionally and creatively. By learning to access and control Theta states, individuals can enhance their mental and emotional well-being, leading to greater self-awareness, emotional intelligence, and creative output.

- **Emotional Intelligence:** Theta waves enable emotional awareness and processing. In a Theta state, the brain can access emotions more directly, allowing individuals to confront and resolve past traumas or negative thought patterns. This emotional clarity can foster stronger emotional intelligence, improving interpersonal relationships and self-regulation.

- **Creative Potential:** The Theta state is a fertile ground for creative breakthroughs. As the brain enters a Theta-dominant state, it is more capable of forming new connections between ideas, making innovative problem-solving easier. Artists, writers, and innovators can use Theta states to enhance their creativity and inspiration, tapping into new ideas and solutions that may have previously been inaccessible.

- **Personal Transformation:** Theta waves facilitate personal transformation by allowing individuals to connect deeply with their inner wisdom and subconscious mind. In this relaxed, open state, individuals can release limiting beliefs, reframe past experiences, and change habitual thought patterns. Many people use Theta meditation for self-discovery, healing, and life changes, unlocking the power to create the life they desire.

## Conclusion

Theta waves are an essential component of the brain's natural rhythm, allowing for deep relaxation, creativity, and emotional healing. Whether you are seeking to unlock your creative potential, access deeper layers of your subconscious mind, or enhance your emotional intelligence, Theta waves provide the key to a richer, more insightful experience of life.

By learning to harness Theta waves through meditation, relaxation, and creative practices, you can foster personal growth, tap into your inner wisdom, and transform your approach to problem-solving and emotional well-being. As we continue our exploration of brainwave mastery, we will see how other frequencies like Delta waves support healing and rest, completing the spectrum of neural oscillations for optimal mental, physical, and emotional health.

# Chapter 7: The Brainwave Spectrum: A Unified View

## Interaction Between Different Brainwave Frequencies

Throughout the previous chapters, we've explored the individual roles of Alpha, Beta, Delta, and Theta waves—each associated with distinct states of consciousness, from focused attention and relaxation to deep sleep and creativity. While each brainwave frequency plays an important role on its own, it's essential to understand that brainwave frequencies don't operate in isolation. Instead, they interact, overlap, and influence each other in a dynamic and interconnected way.

The brain is a highly adaptable and complex system. Rather than being stuck in a single brainwave state, it constantly shifts between different frequencies depending on the mental demands and environment. This constant switching between states is necessary for optimal brain function. Each frequency band serves a specific purpose, and understanding how they interact can provide valuable insights into how we can achieve peak performance, emotional balance, and overall well-being.

For example, during a focused problem-solving session, Beta waves may dominate, signaling active thinking and concentration. However, as the individual takes a break or enters a more relaxed state, Alpha waves might emerge, allowing the brain to rest and refresh before diving back into focused activity. Similarly, creative breakthroughs often occur when Theta waves become more prominent, and the individual can seamlessly shift between states, integrating both logic (Beta) and intuition (Theta).

Understanding how the brain waves interact and transition between these frequencies is vital for mastering the brain's rhythm and enhancing personal performance. The ability to move fluidly from one state to another allows for flexibility in both mental and emotional processing.

## Coherence and Synchronization Between Waves

Coherence refers to the degree to which different brain regions produce synchronized neural oscillations. When the brain is operating in a highly coherent state, different regions communicate efficiently, allowing for smooth, integrated processing of information. In contrast, a lack of coherence can lead to inefficient communication between brain regions, which can manifest as cognitive or emotional dysfunction.

When we consider brainwave synchronization, it's essential to understand that certain mental states require multiple brain regions to be synchronized at different frequencies. For instance, the interaction between Alpha and Theta waves during meditation can produce a state of heightened creativity and calm. Meanwhile, synchronized Beta activity across the prefrontal cortex allows for focused thought and decision-making.

Synchronizing brainwaves across regions enables effective problem-solving, emotional regulation, and cognitive flexibility. For example, during a brainstorming session, the brain may synchronize Theta waves in the creative centers while maintaining Beta waves in the areas responsible for focus and logic. This synchronization allows for the merging of creative intuition and practical problem-solving, leading to breakthroughs.

One key tool for enhancing brainwave coherence is **neurofeedback**, a method that uses real-time brainwave measurements to help individuals train their brains to achieve more synchronized and coherent patterns of brainwave activity. By targeting specific frequencies, neurofeedback can help individuals optimize their brain function, enhance cognitive performance, and achieve better emotional regulation.

## The Importance of Balancing Brainwave States for Optimal Functioning

The ability to balance different brainwave states is essential for optimal functioning, whether in work, creative pursuits, physical activity, or emotional health. Each brainwave frequency has its advantages, but being overly dominant in one frequency can limit your cognitive flexibility or emotional balance.

- **Too Much Beta Activity:** Excessive Beta wave activity can lead to overthinking, stress, anxiety, and mental fatigue. While Beta waves are critical for focused attention, too much activity in this range without adequate Alpha or Delta waves can result in mental exhaustion. This is commonly seen in individuals who work under high pressure for extended periods without taking adequate breaks or practicing relaxation techniques.

- **Too Much Alpha Activity:** While Alpha waves are associated with relaxation, excessive Alpha wave activity can make the mind feel too calm or disengaged, which may hinder productivity or problem-solving in tasks that require active focus. Balancing Alpha with Beta waves ensures that the brain remains alert and engaged while maintaining a relaxed state.

- **Too Much Delta Activity:** While Delta waves are essential for restorative sleep and physical healing, excessive Delta activity during wakefulness can result in brain fog, lethargy, and reduced cognitive performance. Striking the right balance between Delta waves during sleep and other brainwave frequencies during wakefulness is crucial for maintaining energy levels and focus.

- **Too Much Theta Activity:** Theta waves, while powerful for creativity and deep relaxation, can also make the mind overly introspective, leading to disengagement from the present task. A balance of Theta with Beta activity enables creative flow while maintaining practical focus and action.

Balancing brainwave states can be seen as a delicate dance—too much of one frequency can lead to imbalance, while too little can limit performance. It is through the mastery of shifting between different brainwave states that individuals can achieve optimal brain health, cognitive flexibility, and emotional resilience.

## Real–Life Examples of How Brainwaves Work Together

1. **Creativity and Problem-Solving:** During creative problem-solving, the brain may transition between Beta and Theta waves. Theta waves help unlock new ideas and insights, while Beta waves keep the mind focused on bringing those ideas into practical action. This dynamic interaction allows the brain to explore creative solutions while maintaining mental clarity and organization. For instance, an artist might experience Theta-dominant states when coming up with new concepts, while the Beta waves help them refine and bring these ideas to life.

2. **Meditation and Emotional Regulation:** A regular meditation practice often involves shifting between Alpha and Theta waves. As a meditator enters deeper states of relaxation, Alpha waves become dominant, reducing stress and promoting calmness. If the meditation leads to deeper introspection or spiritual insight, Theta waves may become more pronounced, enabling access to the subconscious and deeper emotional processing. The balance of these waves helps regulate emotions, reduce anxiety, and promote overall mental clarity.

3. **Peak Performance in Sports:** Athletes often experience the interaction of multiple brainwave frequencies as they enter a state of "flow." During high-intensity activity, Beta waves dominate, enabling focus, quick reactions, and the execution of learned skills. However, when the athlete enters a relaxed but highly focused state, Alpha waves may emerge, ensuring they stay calm under pressure. In some cases, athletes may also tap into Theta waves to access intuitive insights, such as anticipating a move or reading an opponent's strategy. This synergy of brainwaves allows them to perform at their peak.

4. **Learning and Memory Retention:** Effective learning requires a combination of Beta and Theta waves. Theta waves facilitate memory consolidation and creative problem-solving, while Beta waves help sustain focus and mental effort. For example, when learning a new language, Theta waves support the subconscious processing and integration of new vocabulary, while Beta waves allow the learner to stay focused and organized, ensuring that new information is encoded in working memory.

## Conclusion: The Power of a Unified Brainwave Spectrum

The interaction between different brainwave frequencies is key to understanding how the brain functions at its best. Each frequency, from Alpha to Beta to Theta and Delta, serves a unique purpose, but it is the dynamic integration of these waves that enables optimal cognitive, emotional, and physical performance. Mastering the brainwave spectrum is not about focusing on one frequency alone but about understanding how these waves can work together to support diverse aspects of our lives.

To achieve peak performance, emotional balance, and creative flow, individuals must learn to harness the power of each brainwave frequency and ensure that these states are balanced and synchronized. Practices like neurofeedback, meditation, and brainwave entrainment can help individuals develop a deeper understanding of how to regulate their brainwave activity and optimize their overall well-being.

By cultivating the ability to shift between different brainwave states and synchronize their activity, individuals can unlock their full potential, achieving a state of mental clarity, creativity, and emotional resilience. The brainwave spectrum is a unified system, and by mastering its nuances, you can align your brain with your highest goals and desires, creating a life of balance, growth, and fulfillment.

# Chapter 8: The Relationship Between Brainwaves and Mental Health

## Brainwave Imbalances and Mental Health Disorders

Brainwaves play a fundamental role in regulating our mental states and emotional responses. When brainwave activity is balanced, the brain functions efficiently, promoting cognitive clarity, emotional stability, and optimal health. However, when there is an imbalance or disruption in brainwave patterns, it can have significant impacts on mental health.

Mental health disorders such as anxiety, depression, ADHD, and insomnia have been linked to irregularities in brainwave activity. For example, people with anxiety disorders often show an overabundance of high-frequency Beta waves (greater than 20 Hz), which are associated with heightened alertness and stress. On the other hand, those with depression might experience low Alpha wave activity (8-12 Hz), which can make it harder to relax and maintain a positive mental state. Similarly, people with ADHD often exhibit irregular brainwave patterns, with an underproduction of Theta waves (4-8 Hz) and an overproduction of Beta waves, making it difficult to concentrate or relax.

Understanding how these imbalances manifest and affect behavior is essential for developing strategies to restore healthy brainwave patterns and improve mental well-being.

## How Brainwaves Relate to Anxiety, Depression, and Stress

- **Anxiety and Stress:**

  Anxiety is typically associated with an overactivation of Beta waves (especially in the higher frequencies). Beta waves are linked to focus, concentration, and alertness, but when they dominate excessively, they can cause feelings of agitation, racing thoughts, and mental exhaustion. Chronic stress and anxiety disorders have been shown to produce a pattern of too much Beta wave activity, making it hard to relax or let go of worry.

  By reducing excessive Beta activity and increasing Alpha waves (which promote relaxation), individuals can counterbalance the heightened mental alertness associated with anxiety and create a more relaxed state. Practices like deep breathing, mindfulness meditation, and relaxation techniques are effective at shifting the brain into a more balanced brainwave state, reducing both the physical and mental symptoms of anxiety.

- **Depression:**

  Depression is often marked by a decrease in Alpha wave activity, which can make it difficult for the individual to relax, focus, and feel positive emotions. Low Alpha activity is commonly observed in individuals who struggle to disengage from negative thoughts and feelings, leading to a sense of mental "stuckness" and a lack of motivation. Furthermore, there may also be an imbalance in Theta waves, which are associated with deep emotional processing and introspection. Restoring Alpha wave activity and increasing Theta waves can be instrumental in improving mood and promoting relaxation. Techniques such as mindfulness meditation, which encourage a relaxed and present state, have been shown to help regulate brainwave activity, leading to emotional healing and enhanced well-being.

- **Sleep Disorders:**

  Sleep disorders, including insomnia, are also related to brainwave imbalances. During the deeper stages of sleep, Delta waves (0.5–4 Hz) are required for restorative rest, tissue repair, and memory consolidation. However, people who struggle with insomnia or poor sleep quality often show a lack of Delta wave activity, making it difficult to achieve the deep, restorative stages of sleep. By promoting Delta wave activity—through practices such as meditation, deep breathing, and biofeedback—individuals can improve the quality of their sleep. This helps regulate the circadian rhythm and ensures the body and brain have adequate time to rest and heal. Moreover, incorporating Delta-promoting techniques into the bedtime routine can significantly enhance sleep onset and duration.

## Techniques for Restoring Healthy Brainwave Patterns through Mindfulness and Therapy

Fortunately, there are several effective techniques that can help restore healthy brainwave patterns and improve mental health. These approaches aim to rebalance the frequencies that may be too dominant or too suppressed, helping individuals achieve emotional stability, clarity, and well-being.

## 1. Meditation and Mindfulness Practices

Meditation has long been known for its ability to influence brainwave activity and improve mental health. Regular meditation helps to lower Beta wave activity, reduce stress, and increase Alpha and Theta waves, promoting relaxation, mental clarity, and emotional healing.

- **Mindfulness Meditation:** Mindfulness practices help individuals bring awareness to the present moment, which naturally activates Alpha waves and reduces Beta activity. Mindfulness has been shown to reduce stress, anxiety, and depression by helping individuals break free from rumination and negative thought patterns.

- **Loving-Kindness Meditation (Metta):** This practice encourages positive emotions, such as compassion and love, and has been shown to increase Alpha wave activity while also enhancing emotional well-being. It can be particularly useful in combating feelings of sadness and hopelessness.

- **Transcendental Meditation (TM):** TM involves focusing on a mantra, helping individuals access deep states of relaxation. This technique has been shown to reduce stress and anxiety by increasing Alpha waves and lowering Beta wave activity.

## 2. Neurofeedback

Neurofeedback is a cutting-edge technique that uses real-time monitoring of brainwave activity to help individuals train their brains to regulate brainwave frequencies. This technique involves using EEG to measure brainwave patterns and providing feedback, such as sound or visual cues, when the desired brainwave activity is achieved.

For individuals with anxiety, neurofeedback can be used to reduce excessive Beta wave activity and increase Alpha wave activity. For those with depression, neurofeedback can help raise Alpha waves and improve overall mood. Neurofeedback has also been used effectively to treat ADHD by increasing Theta waves to enhance focus and concentration while reducing Beta waves.

## 3. Brainwave Entrainment

Brainwave entrainment involves using external stimuli, such as sound or light, to synchronize brainwaves to a desired frequency. Binaural beats, isochronic tones, and light frequencies are all commonly used methods for brainwave entrainment. By listening to audio tracks or engaging with light frequencies that target specific brainwave bands, individuals can facilitate changes in brainwave activity.

- **Binaural Beats:** Binaural beats are created by presenting two slightly different frequencies in each ear, which the brain perceives as a single, new frequency. For instance, listening to binaural beats designed to increase Alpha waves can help calm anxiety and promote relaxation.
- **Isochronic Tones:** These are single tones that pulse at a specific frequency to entrain the brain. They are often used for therapeutic purposes and can be tailored to address conditions such as stress, anxiety, or sleep disturbances by promoting the production of specific brainwave frequencies.

## 4. Cognitive Behavioral Therapy (CBT) and Other Psychotherapies

Psychotherapies like Cognitive Behavioral Therapy (CBT) can help address the thought patterns that contribute to mental health disorders. While CBT does not directly alter brainwave activity, it can lead to changes in how the brain processes thoughts and emotions, potentially influencing brainwave patterns over time.

For example, CBT can help reduce negative thought patterns that lead to excessive Beta wave activity (such as excessive worry) and increase awareness of more positive, present-moment thinking (which is linked to Alpha and Theta waves). By incorporating mindfulness techniques into therapy, patients can further promote healthy brainwave regulation and emotional balance.

## Using Brainwave Training to Address Specific Mental Health Conditions

**Anxiety and Stress:**

- Use **Alpha-enhancing** techniques like mindfulness and binaural beats to promote relaxation and reduce excessive Beta wave activity.
- Practice **deep breathing** and **guided imagery** to shift the brain from a stressed, Beta-dominant state into a calmer, more balanced state.

## Depression:

- Engage in practices that **increase Alpha and Theta waves**, such as meditation or neurofeedback, to enhance emotional healing and break negative thought patterns.
- Use **brainwave entrainment** to promote a more positive mood and greater mental clarity.

## Sleep Disorders:

- Use **Delta-promoting** techniques, such as meditation, to foster deep, restorative sleep.
- Try **binaural beats** specifically designed to promote Delta waves and enhance the quality of sleep.

## ADHD:

- Engage in neurofeedback or **Theta-enhancing techniques** to improve focus and attention.

- Use **mindfulness and cognitive exercises** to regulate brain activity and help increase self-awareness.

## Conclusion

Brainwave imbalances are closely tied to mental health disorders, but through mindful techniques and emerging technologies, it is possible to restore balance and improve well-being. By understanding the relationship between brainwaves and mental health, individuals can take active steps to optimize their brain function, alleviate symptoms of anxiety, depression, and stress, and create a more harmonious mental state.

In the next chapters, we will explore tools such as neurofeedback and brainwave entrainment to better understand how we can actively use these methods to fine-tune brainwave activity and enhance mental health. By mastering our brainwaves, we can unlock the potential for emotional healing, mental clarity, and overall well-being.

# Chapter 9: Measuring Brainwaves: Tools and Techniques

## Overview of EEG and Other Brainwave Measurement Devices

Brainwaves, though invisible to the naked eye, can be measured using specialized tools that capture the electrical activity of the brain. One of the most widely used devices for this purpose is the **electroencephalogram (EEG)**. EEG is a non-invasive technique that records the electrical activity of the brain by placing electrodes on the scalp. The EEG detects the small electrical signals produced when neurons communicate, allowing us to measure and analyze the various brainwave frequencies.

EEG has been a critical tool in neuroscience and clinical practice for decades. It helps researchers and clinicians understand the brain's electrical activity, diagnose neurological conditions like epilepsy, and assess cognitive and emotional states. Modern EEG systems provide real-time data on brainwave patterns, offering detailed insights into how brainwave frequencies fluctuate in response to different mental tasks, emotional states, and health conditions.

EEG provides a visual representation of brainwaves as waveforms, which can be classified into the primary frequency bands: Delta, Theta, Alpha, Beta, and Gamma. The EEG can reveal which brainwave frequencies dominate during certain activities—whether a person is engaged in deep relaxation, focused concentration, or in deep sleep. In clinical settings, EEG can help monitor brain function, detect abnormalities, and even assess the effectiveness of therapeutic interventions.

In addition to EEG, there are several other tools and devices that help measure brain activity and facilitate brainwave analysis. These include:

- **Functional Magnetic Resonance Imaging (fMRI):** While not a direct measure of brainwaves, fMRI measures changes in blood flow and can indicate brain activity. It is often used in conjunction with EEG for more comprehensive brain studies.
- **Magnetoencephalography (MEG):** MEG detects the magnetic fields produced by neural activity and provides a detailed map of brain activity in real-time. It is particularly useful in identifying the source and timing of brainwave patterns.
- **Near-Infrared Spectroscopy (NIRS):** NIRS is a method used to measure brain oxygenation, which can correlate with brainwave activity. While it doesn't directly measure brainwaves, it can provide valuable complementary data for understanding brain function.

## Neurofeedback and Its Applications

Neurofeedback, also known as EEG biofeedback, is a therapeutic technique that trains individuals to alter their brainwave activity in real-time by providing immediate feedback on their brainwave patterns. The goal of neurofeedback is to help individuals learn how to regulate their brainwaves, improving mental health and cognitive performance.

In neurofeedback sessions, EEG sensors are placed on the scalp to measure brainwave activity. The feedback is typically given through auditory, visual, or tactile signals, which inform the individual about their brainwave patterns. If a person's brainwaves are in the desired range (for example, increasing Alpha waves or reducing excessive Beta waves), the feedback might be a rewarding sound or visual cue. Conversely, if the brainwaves fall outside the desired range, the feedback might alert the individual to adjust their state.

Neurofeedback is commonly used to treat a variety of conditions, including:

- **ADHD (Attention-Deficit/Hyperactivity Disorder):** Many people with ADHD have an overproduction of Beta waves (associated with high mental activity) and insufficient Theta waves (linked to calmness and focus). Neurofeedback can help individuals learn to increase Theta waves while decreasing excessive Beta waves, improving attention and impulse control.

- **Anxiety and Stress:** By training individuals to increase Alpha waves (associated with relaxation) and decrease Beta waves (linked to stress and anxiety), neurofeedback helps individuals manage symptoms of anxiety, promoting relaxation and emotional regulation.

- **Sleep Disorders:** Neurofeedback can also help individuals improve their sleep quality by encouraging Delta waves (associated with deep sleep) and regulating other brainwave frequencies that disrupt the sleep cycle.

- **Depression:** Studies have shown that neurofeedback can be effective for treating depression by balancing brainwave activity, particularly by increasing Alpha waves and reducing excessive Beta wave activity, which is often seen in individuals with depression.

Neurofeedback is a powerful tool for self-regulation, offering a non-invasive way to enhance brain function and improve mental health. It provides real-time feedback on brainwave patterns, enabling individuals to become more aware of their mental states and learn how to modify them for improved well-being.

## The Growing Field of Brain-Computer Interfaces (BCIs)

Brain-Computer Interfaces (BCIs) are a rapidly emerging technology that directly connects the brain to external devices. BCIs can record brain activity, decode it into actionable signals, and use those signals to control devices or provide feedback to the user. This technology has far-reaching potential applications in both medical and non-medical fields, ranging from helping individuals with paralysis control prosthetics to enhancing cognitive performance in healthy individuals.

In the context of brainwave mastery, BCIs have opened up exciting possibilities for real-time brainwave manipulation and optimization. For example, BCI systems can be used to monitor and alter brainwave patterns, providing users with precise control over their mental states. These systems can be especially useful in training and therapeutic settings, where individuals aim to enhance mental focus, creativity, relaxation, or emotional regulation.

Some practical applications of BCIs in brainwave training include:

- **Mental Training and Cognitive Enhancement:** BCIs can assist in cognitive training by providing real-time feedback on brain activity, helping individuals optimize their brainwaves for tasks such as memory improvement, enhanced focus, and problem-solving.

- **Rehabilitation:** For individuals recovering from brain injuries, BCIs can be used to facilitate neuroplasticity by guiding the brain to establish new connections. They can help individuals regain motor control, improve cognitive functions, or reduce the impact of neurological disorders.

- **Augmenting Performance in Athletes and Artists:** BCIs have been used in sports psychology to train athletes to achieve peak mental states. By using BCIs to monitor and adjust brainwave activity, athletes can enhance focus, reduce performance anxiety, and optimize their mental state for high-stakes events.

- **Mental Health Treatment:** BCIs also have potential applications in treating mental health conditions. By using BCIs to monitor brain activity, therapists can guide patients in self-regulating their brainwaves to alleviate symptoms of anxiety, depression, and PTSD.

While still a developing field, BCIs hold great promise for enhancing cognitive abilities, improving mental health, and even unlocking new forms of human-computer interaction.

## Understanding Your Own Brainwave Patterns with Technology

As individuals become more aware of the impact of brainwaves on mental health, cognitive performance, and emotional well-being, there is an increasing interest in understanding and regulating one's own brainwave patterns. Fortunately, advancements in technology have made it easier for individuals to measure and track their brainwaves outside of clinical settings.

Several consumer-grade EEG headsets are now available on the market, allowing people to monitor their brainwave activity at home. These devices typically connect to smartphones or computers, providing feedback on brainwave patterns in real-time. Some popular brands include:

- **Muse:** The Muse headband is a popular EEG device that tracks brainwave activity and provides feedback during meditation. It helps users improve focus, calmness, and relaxation by guiding them through a series of mental exercises.
- **Emotiv:** Emotiv offers EEG devices that are used for both brainwave measurement and cognitive enhancement. Their systems are used for everything from stress management to improving productivity and creativity.
- **NeuroSky:** NeuroSky is another brand that produces affordable, consumer-grade EEG headsets that provide real-time feedback on brainwave activity. These devices are often used in cognitive training and biofeedback applications.

These devices offer individuals the opportunity to track their brainwave patterns over time, helping them understand how their mental states change in response to different activities, moods, and experiences. By incorporating these devices into daily life, individuals can learn to control their brainwaves more effectively, enhancing their overall mental and emotional well-being.

## Conclusion

The ability to measure and manipulate brainwaves has evolved dramatically in recent years, thanks to advances in EEG technology, neurofeedback, and brain-computer interfaces. These tools provide valuable insights into the functioning of the brain, enabling individuals to better understand their mental states and develop strategies for optimizing cognitive performance, emotional regulation, and overall well-being.

From EEG devices that track brainwave patterns in real-time to the emerging field of BCIs, these tools offer tremendous potential for enhancing mental health, improving performance, and even unlocking the brain's full creative potential. By harnessing the power of these tools, individuals can take control of their brainwave activity, achieve peak mental performance, and live a more balanced, fulfilling life.

In the next chapters, we will explore how brainwave entrainment and other techniques can be used to train the brain, further enhancing cognitive function and personal growth. With a deeper understanding of how to measure and influence brainwaves, you can begin your journey towards mastering neural oscillations and optimizing your brain for success.

# Chapter 10: Enhancing Cognitive Performance with Brainwaves

### How Mastering Brainwaves Can Improve Memory, Focus, and Learning

The brain is an extraordinary organ capable of remarkable cognitive feats. However, achieving peak cognitive performance requires more than just raw intelligence or willpower. It depends on optimizing the brain's natural rhythms and understanding how to leverage brainwaves for specific mental tasks. Mastering brainwave frequencies can significantly enhance key cognitive functions like memory, focus, and learning— facilitating better academic, professional, and personal performance.

Each brainwave frequency plays a role in cognitive functions, and by deliberately regulating brainwave activity, individuals can improve specific aspects of their mental capabilities. For example, by boosting Alpha wave activity, you can enter a relaxed but alert state, which is ideal for creativity and memory retention. Conversely, Beta waves are important for concentrated thought, and increasing them can improve focus and productivity, particularly when performing complex tasks.

When cognitive performance is optimized by brainwave mastery, learning becomes more efficient, memory recall improves, and mental clarity is enhanced. The ability to control brainwave activity also leads to better emotional regulation, which further supports cognitive function. Here's a deeper look at how mastering brainwaves impacts each cognitive function:

- **Memory:** Alpha and Theta waves are key players in enhancing memory consolidation. Alpha waves support the integration of short-term information into long-term memory, while Theta waves facilitate deeper memory retrieval and creativity. By training the brain to boost these waves, you can improve memory retention and recall, making learning faster and more effective.

- **Focus and Attention:** Beta waves are crucial for sustained attention and concentration. Mastering Beta waves helps to improve focus during mentally demanding tasks. When Beta waves are regulated, you can maintain high levels of concentration for longer periods of time without experiencing mental fatigue.

- **Learning:** Theta waves support creativity and learning by facilitating access to subconscious thoughts and enhancing problem-solving. By learning to induce and control Theta states, you can tap into intuitive insights and make connections between seemingly unrelated ideas, accelerating the learning process.

By understanding and mastering the relationship between brainwave frequencies and cognitive functions, you can unlock higher levels of performance and productivity. This allows you to learn faster, concentrate better, and recall information more effectively.

## Using Brainwave Training to Improve IQ and Cognitive Speed

Brainwave training is a proven method for enhancing cognitive performance. This training works by helping individuals deliberately modify their brainwave patterns to align with the frequencies associated with optimal mental states. One of the main applications of brainwave training is improving cognitive speed and IQ.

- **Cognitive Speed:** The ability to think quickly and process information efficiently is highly valued in today's fast-paced world. Beta wave activity is associated with high-level cognitive processing and fast thinking. By using neurofeedback or other brainwave training techniques to increase Beta waves, individuals can improve their cognitive speed, allowing them to analyze information, make decisions, and solve problems faster.

- **Improving IQ:** IQ is often associated with an individual's capacity for logical reasoning, problem-solving, and abstract thinking—all processes that require synchronization between multiple brain regions. Brainwave training can help optimize the brain's frequency patterns, promoting enhanced cognitive abilities. By stimulating specific brainwave frequencies such as Beta and Theta, individuals can improve their problem-solving abilities and intellectual capacity.

Brainwave training techniques like **neurofeedback** and **brainwave entrainment** can help individuals target specific brainwave states that are conducive to faster cognitive processing and higher IQ. Neurofeedback helps individuals learn to control their brainwaves by providing real-time feedback, allowing them to adjust their brainwave patterns consciously. With consistent practice, the brain becomes more efficient, leading to improvements in IQ and cognitive speed.

## Practical Applications of Brainwave Entrainment for Students and Professionals

Brainwave entrainment is a technique that uses external stimuli—such as sounds, light, or rhythmic pulses—to synchronize brainwaves to a desired frequency. This powerful tool can help individuals optimize their cognitive performance, whether they are students studying for exams, professionals aiming to boost productivity, or athletes preparing for competition.

Here's how brainwave entrainment can be applied in various settings:

- **For Students:**

  Brainwave entrainment can be used to improve focus, memory retention, and learning efficiency. By listening to binaural beats or isochronic tones tuned to specific brainwave frequencies (Alpha or Theta waves), students can enter an optimal learning state, making it easier to absorb new information and recall it during exams.

  Studies have shown that brainwave entrainment can enhance memory consolidation by promoting the necessary brainwave patterns for information storage. For example, listening to Theta or Alpha waves before studying can help improve memory recall, while Beta wave entrainment can enhance focus and concentration during study sessions.

- **For Professionals:**

  Professionals who need to perform high-concentration tasks can use brainwave entrainment to boost productivity. Beta wave entrainment is particularly effective for increasing focus and mental clarity, allowing individuals to work more efficiently. It can also help reduce mental fatigue, as it keeps the brain in an optimal state for sustained cognitive effort.

  For those in high-stress careers, Alpha wave entrainment can be used to reduce stress and promote relaxation, while still maintaining mental clarity. This can enhance both performance and well-being, allowing professionals to manage high workloads and deadlines without burnout.

- **For Athletes:**

   Athletes can benefit from brainwave entrainment by using it to improve focus, reaction times, and mental preparation before competitions. Alpha waves help athletes relax while remaining mentally sharp, while Beta waves enhance focus and concentration during intense activity. Theta waves can also help athletes access deeper levels of creativity and intuitive problem-solving, which can be particularly useful in strategy-based sports.

   Athletes can use brainwave entrainment during training to ensure that they remain calm and focused under pressure, enhancing their performance during competition. The ability to switch between Alpha, Beta, and Theta waves provides athletes with greater control over their mental state and performance.

## Brainwave Training Techniques for Cognitive Enhancement

There are several practical methods for training the brain and optimizing brainwave activity. These methods allow individuals to enhance their cognitive abilities and improve mental performance over time. Here are a few examples:

- **Neurofeedback Training:**

  Neurofeedback is one of the most powerful tools for brainwave training. This technique involves using sensors to measure brainwave activity in real time and providing feedback to help individuals learn to regulate their brainwaves. Through neurofeedback, individuals can train their brains to increase or decrease specific brainwave frequencies, leading to improved focus, memory, and cognitive processing.

- **Binaural Beats and Isochronic Tones:**

  These auditory tools help entrain brainwaves to specific frequencies. Binaural beats use two slightly different frequencies played in each ear to produce a third frequency in the brain, while isochronic tones use rhythmic pulses to synchronize brainwaves. Both methods are effective for guiding the brain into desired states, whether for relaxation (Alpha), creativity (Theta), or focus (Beta).

- **Meditation and Mindfulness Practices:**

  Meditation is a time-tested method for enhancing brain function and promoting mental clarity. Regular meditation practice can increase Alpha and Theta waves, improving cognitive performance, creativity, and emotional regulation. Mindfulness practices, such as deep breathing and body scans, help individuals relax and focus their attention, which in turn enhances cognitive abilities.

- **Cognitive Training Exercises:**

  Engaging in cognitive exercises, such as memory games, puzzles, and brain training apps, can help improve cognitive speed and mental agility. These activities help strengthen the brain's ability to process information quickly and efficiently, boosting IQ and overall cognitive performance.

## Conclusion

Mastering brainwave activity is a powerful way to enhance cognitive performance, boost memory, improve focus, and increase learning efficiency. By understanding the connection between brainwaves and cognitive functions, individuals can harness the power of their brain's natural rhythms to achieve peak mental performance.

Brainwave training, whether through neurofeedback, brainwave entrainment, or meditation, provides a valuable toolkit for improving cognitive abilities and emotional regulation. Whether you are a student looking to optimize learning, a professional striving for peak productivity, or an athlete aiming to enhance performance, mastering brainwaves can help you achieve your goals with greater ease and efficiency.

In the following chapters, we will explore more advanced brainwave control techniques and delve deeper into how these practices can be applied in various aspects of life to unlock greater mental, emotional, and physical potential.

# Chapter 11: Brainwave Entrainment: The Power of Sound and Light

## Understanding Brainwave Entrainment

Brainwave entrainment refers to the process of synchronizing brainwave frequencies with an external stimulus, such as sound or light. This synchronization helps the brain shift into specific states conducive to relaxation, focus, creativity, or sleep. The underlying principle behind brainwave entrainment is that the brain is highly responsive to rhythmic stimuli and naturally aligns its brainwave activity with the frequency of these stimuli.

The most common forms of brainwave entrainment involve **auditory** and **visual** cues. These external stimuli, when presented at specific frequencies, can guide the brain to adopt the corresponding brainwave state. This process of entrainment can be highly effective in promoting desired mental states and has been applied in various contexts, from reducing stress and anxiety to improving cognitive performance and creativity.

## Binaural Beats, Isochronic Tones, and Light Frequencies
## Binaural Beats

Binaural beats are a form of auditory brainwave entrainment that uses two distinct sound frequencies played separately into each ear. The brain perceives a third frequency—the difference between the two tones—and aligns its electrical activity to match this new frequency. For example, if a 300 Hz tone is played in one ear and a 310 Hz tone is played in the other, the brain perceives a 10 Hz beat.

- **Application of Binaural Beats:**

  Binaural beats are often used to promote specific brainwave frequencies, such as Alpha (8-12 Hz) for relaxation and focus, Theta (4-8 Hz) for creativity and deep meditation, or Delta (0.5-4 Hz) for deep sleep. By using binaural beats with targeted frequencies, individuals can induce states of relaxation, enhance cognitive performance, or improve sleep quality.

- **Benefits:**

  Research has shown that binaural beats can help reduce anxiety, improve focus, enhance memory retention, and even promote restful sleep. Many people use binaural beats during meditation or study sessions to improve concentration and mental clarity.

## Isochronic Tones

Isochronic tones are single tones that pulse at specific intervals to create a rhythmic pattern that the brain can synchronize with. Unlike binaural beats, which require headphones, isochronic tones can be heard with or without stereo sound, making them more versatile. The pulsing tones are designed to resonate at a specific frequency that corresponds to the desired brainwave state.

- **Application of Isochronic Tones:**
  Isochronic tones can be used to entrain the brain to various frequencies, such as Alpha, Beta, Theta, or Delta waves. These tones are typically used for promoting relaxation, enhancing mental performance, or facilitating deeper sleep.

- **Benefits:**
  Isochronic tones have been shown to reduce stress, improve attention, enhance creativity, and optimize sleep. The rhythmic pulsing of the tones stimulates brainwave activity in a way that is simple and effective, making them a popular tool for cognitive enhancement and mental health.

## Light Frequencies (Photobiomodulation)

In addition to auditory brainwave entrainment, visual stimuli—such as light frequencies—can also influence brainwave patterns. This approach is based on the idea that light can stimulate the brain to produce specific brainwave frequencies through exposure to pulsed light.

- **Application of Light Frequencies:**

  Light frequencies can be delivered through devices like light glasses or LED-based systems. The light pulses at specific frequencies, guiding the brain to adopt certain brainwave states. For example, using light to promote Alpha or Theta wave activity can encourage relaxation or deep meditation, while higher frequencies can support alertness and focus.

- **Benefits:**

  Light-based brainwave entrainment can be effective for managing sleep cycles, reducing symptoms of seasonal affective disorder (SAD), enhancing concentration, and even boosting mood. The use of light frequencies offers a non-invasive and convenient option for altering brainwave states.

## How Sound and Light Influence Brainwave Patterns

The brain responds to rhythmic stimuli in a way that aligns its oscillatory patterns with the frequency of the external cues. When presented with sound or light that pulses at a specific frequency, the brain synchronizes its electrical activity with this rhythm, a phenomenon known as **frequency-following response (FFR)**.

The brain is particularly responsive to external rhythmic patterns, as they provide a predictable signal that the brain can easily follow. This is why rhythmic sound and light pulses are effective in guiding the brain to a specific state. By providing the brain with a consistent and regular stimulus, entrainment helps optimize mental functioning for various activities, whether it's relaxation, focus, creativity, or sleep.

- **Relaxation and Stress Relief:**

  Alpha and Theta frequencies, for example, are linked to relaxation and calmness. When you listen to binaural beats or isochronic tones that promote these frequencies, the brain enters a relaxed state, reducing stress and anxiety. The relaxation response can help regulate the autonomic nervous system, lowering heart rate and blood pressure, and promoting overall emotional well-being.

- **Enhanced Focus and Productivity:**

  Beta frequencies, associated with active thinking and problem-solving, can be induced using auditory or visual entrainment. By stimulating Beta waves, you can achieve a state of heightened alertness and concentration. This can be particularly useful when studying, working, or engaging in tasks that require sustained attention.

- **Improved Sleep:**

  Delta waves, linked to deep restorative sleep, can be stimulated using brainwave entrainment techniques. Listening to Delta-promoting binaural beats or using light frequencies that encourage Delta wave production can help improve the quality of sleep, aiding in the body's healing and memory consolidation processes.

## Using Brainwave Entrainment for Relaxation and Focus

Brainwave entrainment offers a convenient and effective tool for managing stress, improving focus, and enhancing cognitive function. Below are some practical applications of brainwave entrainment that can benefit different aspects of life:

- **For Stress Relief and Relaxation:**

  Using **Alpha**-promoting binaural beats or isochronic tones can help you achieve a relaxed state and reduce anxiety. Listening to these frequencies during your break or before bed can help ease tension and encourage a peaceful state of mind.

- **For Enhanced Creativity and Meditation:**

  **Theta** waves, which are associated with deep meditation, creativity, and subconscious access, can be induced through theta-promoting sound frequencies. Using Theta entrainment techniques while meditating or brainstorming can help unlock creative ideas and insights.

- **For Better Focus and Mental Clarity:**

  To improve concentration and task performance, **Beta**-inducing brainwave entrainment is effective. Listening to Beta frequencies while working, studying, or engaging in tasks that require sustained focus can help maintain mental clarity and productivity.

- **For Sleep Optimization:**

  If you struggle with sleep, using **Delta**-stimulating brainwave entrainment techniques before bedtime can help promote deep, restorative sleep. Delta waves are essential for the body's healing and repair processes, and using entrainment to increase Delta activity can significantly improve sleep quality.

## Practical Tools for Brainwave Entrainment

Several tools are available to help individuals engage in brainwave entrainment and optimize their brainwave activity. These tools include:

- **Binaural Beat Apps and Devices:** There are many mobile apps and audio devices that offer binaural beats for different brainwave frequencies. Apps like **Brain.fm**, **Calm**, or **Relax Melodies** provide audio tracks specifically designed to enhance relaxation, focus, or creativity using brainwave entrainment.

- **Isochronic Tone Players:** Similar to binaural beat apps, there are dedicated devices and software that offer isochronic tones designed to induce specific brainwave frequencies. These tools are often used for relaxation, focus, or sleep improvement.

- **Light-Based Entrainment Devices:** Light glasses and light-based systems, such as those offered by **BrainLight** or **Luma**, use light pulses to induce specific brainwave patterns. These can be used to promote relaxation or focus, depending on the frequency of the light pulse.

## Conclusion

Brainwave entrainment through sound and light offers a powerful tool for enhancing mental well-being, improving cognitive performance, and optimizing overall brain health. By synchronizing the brain with specific frequencies, entrainment helps regulate mental states and encourages desired outcomes, such as relaxation, focus, creativity, and sleep. With the growing range of devices and technologies available, individuals can easily incorporate brainwave entrainment into their daily routines to improve mental clarity, emotional balance, and cognitive function.

In the next chapters, we will explore the role of Alpha waves in stress reduction, how Beta waves contribute to peak performance, and how Theta and Delta waves can promote creativity and healing. These techniques, combined with brainwave entrainment, offer a comprehensive approach to mastering neural oscillations and unlocking your brain's full potential.

# Chapter 12: The Role of Alpha Waves in Stress Reduction

### Alpha Waves as a Stress-Relieving Tool

Alpha waves, typically in the frequency range of 8–12 Hz, are often associated with a relaxed yet alert state of mind. These brainwaves are most prominent when we are calm, focused, and in a meditative or light concentration state. Research has shown that Alpha waves play a critical role in stress reduction by promoting relaxation without causing drowsiness. This state is the mental sweet spot where the body experiences reduced stress levels, the mind remains active and clear, and cognitive functions are optimized.

One of the reasons Alpha waves are so effective for stress reduction is that they help synchronize the brain's hemispheres, promoting balanced activity across both sides of the brain. When brainwave patterns are synchronized, the nervous system enters a more relaxed state, leading to reduced cortisol levels (the primary stress hormone). This can directly alleviate feelings of anxiety, promote a sense of calm, and prevent the long-term health issues that arise from chronic stress.

Alpha waves provide a powerful antidote to the chaotic and overstimulated mental state that often accompanies stress. By promoting a state of relaxed alertness, they allow the brain to shift away from the fight-or-flight response (associated with Beta waves) and toward a more balanced, relaxed mode that is conducive to creativity, problem-solving, and well-being.

## Using Alpha Waves to Manage High-Stress Situations

In today's fast-paced world, stress is an unavoidable part of daily life. Whether it's a looming deadline, a high-stakes presentation, or personal challenges, high-stress situations are a common experience. Alpha waves provide a powerful tool to help individuals manage stress by inducing a state of relaxation and mental clarity, even in the face of pressure.

Here are some practical ways to use Alpha waves to manage high-stress situations:

- **Mindfulness and Meditation:**

  Mindfulness meditation is one of the most effective ways to stimulate Alpha wave activity. By focusing on your breath, practicing body scans, or simply observing your thoughts without judgment, you can shift your brain into an Alpha state. This not only reduces stress but also increases awareness and presence, allowing you to navigate stressful situations with a clear and focused mind. Techniques like **guided meditation** (especially those aimed at calming the mind or managing anxiety) are also beneficial for boosting Alpha wave activity. Listening to audio guides or using brainwave entrainment tools can help you enter a relaxed state quickly, even in stressful environments.

- **Visualization:**

  Visualization exercises are another effective method for generating Alpha waves. By imagining a peaceful environment, such as a calm beach or a serene forest, you activate the brain's Alpha activity. This creates a mental break from stressors and helps the body relax. A short visualization session during stressful moments —like before a meeting or a difficult conversation—can reduce tension and improve performance.

- **Breathing Techniques:**

  Deep breathing exercises, especially those that focus on slow, controlled inhales and exhales, can stimulate Alpha wave activity. Slow breathing reduces the fight-or-flight response and enhances the parasympathetic nervous system (the "rest and digest" system). Techniques such as **4-7-8 breathing** or **diaphragmatic breathing** (breathing from the diaphragm rather than shallow chest breathing) are effective ways to calm the mind and promote Alpha waves.

- **Progressive Muscle Relaxation (PMR):**

  Progressive muscle relaxation is a technique where you systematically tense and then relax different muscle groups in the body. This practice helps to release physical tension associated with stress while also promoting a state of mental relaxation. As you relax the muscles, Alpha waves increase, helping to reduce anxiety and improve emotional regulation.

- **Biofeedback Training:**

  Biofeedback is a technique that provides real-time data on your physiological processes, such as heart rate, skin temperature, and brainwave activity. By using biofeedback, you can learn to consciously regulate your brainwave patterns, including enhancing Alpha wave activity. This technique is particularly useful for individuals who experience chronic stress or anxiety, as it helps them gain greater control over their stress response.

Several studies and real-life case studies illustrate the effectiveness of Alpha waves in reducing stress and enhancing mental health. Let's explore some examples:

## Case Study 1: Alpha Waves for Stress Reduction in the Workplace

In a study published in the journal *Neuropsychologia*, researchers explored the impact of Alpha wave training on stress levels in office workers. Participants who practiced mindfulness meditation, designed to enhance Alpha wave production, reported a significant decrease in stress after just a few weeks of practice. They also showed improved cognitive function and productivity, demonstrating that inducing Alpha waves can not only reduce stress but also boost performance in high-pressure work environments.

The study also highlighted that workers who regularly engaged in mindfulness and meditation reported higher job satisfaction and better emotional regulation, emphasizing the broader benefits of Alpha wave enhancement for well-being.

## Case Study 2: Alpha Waves in Anxiety Treatment

Another study published in *Psychiatry Research: Neuroimaging* examined the effects of Alpha wave stimulation on individuals with generalized anxiety disorder (GAD). Participants who underwent neurofeedback training to increase their Alpha wave activity showed a marked reduction in anxiety symptoms. After several sessions, they also reported improved mood and a greater sense of control over their thoughts and emotions.

This study suggests that increasing Alpha wave activity can be a valuable adjunct to traditional therapeutic approaches for managing anxiety. By promoting a more relaxed and balanced mental state, individuals with anxiety disorders can experience greater emotional resilience and fewer panic episodes.

## Case Study 3: Relaxation and Alpha Waves in Students

In a study involving university students preparing for exams, researchers found that listening to Alpha wave-enhancing binaural beats before studying improved both their test performance and stress levels. Students who used the binaural beats showed a significant reduction in exam-related anxiety, along with enhanced memory retention and cognitive performance.

This case study underscores the value of Alpha waves for managing stress in high-pressure situations, such as academic exams. Students who engaged in Alpha-boosting techniques felt more relaxed and mentally clear, leading to improved performance under pressure.

## Alpha Waves and Long-Term Stress Management

While Alpha waves are effective in managing immediate stress, they also have long-term benefits for chronic stress reduction. Regular practice of activities that promote Alpha wave activity can help rewire the brain's stress response, making it more resilient to future stressors. Over time, individuals who regularly engage in relaxation techniques like meditation, breathing exercises, and visualization experience lower levels of overall stress, enhanced emotional well-being, and improved quality of life.

- **Neuroplasticity and Alpha Waves:**

  Neuroplasticity is the brain's ability to reorganize and form new neural connections. Studies have shown that regular mindfulness meditation, which enhances Alpha waves, can foster positive changes in the brain's structure, particularly in areas related to emotional regulation and stress response. This means that over time, engaging in Alpha-promoting practices can help the brain become more adept at managing stress and improving overall mental health.

- **Stress Resilience:**

  One of the long-term benefits of regular Alpha wave stimulation is the improvement in stress resilience. When the brain is trained to regularly enter relaxed states through Alpha wave enhancement, it becomes better equipped to handle stress in everyday life. This enhanced resilience is not only beneficial for mental health but also for physical health, as chronic stress is linked to a variety of conditions such as heart disease, insomnia, and weakened immune function.

## Conclusion

Alpha waves are a powerful tool in managing stress and promoting a sense of calm and focus. By intentionally increasing Alpha wave activity, individuals can reduce anxiety, enhance emotional regulation, and improve cognitive performance. Whether through mindfulness practices, neurofeedback, breathing exercises, or biofeedback, Alpha waves provide a natural and effective way to restore balance in the brain and alleviate stress.

The evidence supporting the role of Alpha waves in stress reduction is compelling, with numerous studies and case studies highlighting their effectiveness. As individuals learn to harness the power of Alpha waves, they can experience greater emotional resilience, improved mental clarity, and a more peaceful approach to life's challenges.

In the next chapter, we will explore the role of Beta waves in achieving peak performance, further expanding on how brainwave mastery can improve not just stress management, but also overall success in various areas of life.

# Chapter 13: Beta Waves and Peak Performance

### Utilizing Beta Waves for Peak Performance in Sports and Business

Beta waves, which occur in the frequency range of 12 to 30 Hz, are associated with active thinking, concentration, and heightened alertness. These brainwaves are present when we are focused on problem-solving, decision-making, and engaging in tasks that require mental effort. While Beta waves are crucial for peak performance in many cognitive tasks, too much of this frequency can lead to stress, mental fatigue, and burnout. Therefore, understanding how to regulate Beta waves is essential for maintaining optimal performance without overwhelming the brain.

In high-pressure environments such as sports and business, Beta waves play a key role in achieving peak performance. Athletes, for example, need Beta waves for rapid decision-making and quick reactions during competition. Similarly, professionals in fast-paced work environments require Beta waves to stay focused on complex tasks, manage multiple projects, and solve problems efficiently.

However, the challenge lies in maintaining an optimal level of Beta activity. Too much Beta wave activity can lead to overstimulation, stress, and burnout, while insufficient Beta activity can result in poor focus and indecisiveness. The goal is to balance Beta activity in a way that enhances performance while preventing stress overload.

## Beta Waves in Sports Performance

In the world of sports, athletes must make fast, split-second decisions, execute complex motor skills, and maintain a high level of focus. Beta waves are key to this type of high-intensity performance. These brainwaves help athletes stay alert, react quickly, and perform tasks with precision.

- **Optimal Beta Waves for Quick Decision Making:**

  In fast-paced sports like basketball or soccer, athletes must constantly evaluate the environment and make rapid decisions. This is where Beta waves shine. By increasing Beta wave activity, athletes can improve their ability to process information quickly and respond appropriately. Beta waves enhance reaction times, ensuring that athletes are mentally sharp and prepared for every move.

- **Focus and Mental Clarity:**

  Maintaining a sharp focus is essential for athletes. Beta waves help keep the mind focused on the task at hand, whether it's executing a complicated gymnastic move or tracking the ball in tennis. By tuning Beta waves, athletes can block out distractions and stay mentally present during competition, ensuring they perform at their best.

- **Handling Stress and Anxiety:**

  While Beta waves are associated with focus, too much Beta activity can lead to performance anxiety and stress. This is often seen in athletes before important competitions. Managing Beta waves becomes crucial in these situations. Techniques like deep breathing, meditation, and mindfulness can help reduce excessive Beta activity, allowing athletes to remain calm and focused under pressure.

Beta Waves in Business and Professional Settings

In business, particularly in high-stakes situations such as meetings, presentations, or negotiations, maintaining peak mental performance is essential. Beta waves are critical for cognitive processing, creative problem-solving, and sustained attention to detail. However, just like in sports, the challenge is balancing Beta wave activity to avoid cognitive overload.

While Beta waves are essential for peak performance, the key to optimal functioning is maintaining a balanced level of Beta activity. Too much Beta activity can lead to mental exhaustion and stress, while too little can cause a lack of focus and sluggish thinking. Here are a few techniques to ensure that Beta waves support peak performance without overloading the brain:

## 1. Mindfulness and Meditation

Mindfulness and meditation techniques are valuable tools for managing Beta wave activity. By practicing mindfulness, you can enhance the beneficial aspects of Beta waves—such as focus and concentration—while reducing the negative effects of overstimulation, like anxiety and stress.

- **Focused Attention Meditation:**

  Focusing on a single point, whether it's your breath or a visual object, helps cultivate Beta waves that are associated with attention. This technique helps you increase your ability to focus and sustain attention without overwhelming the mind. Over time, you can train yourself to maintain this state of focused concentration in high-pressure situations, such as during meetings or competitions.

- **Body Scan Meditation:**

  This technique involves mentally scanning your body for tension and consciously releasing it. A body scan meditation helps reduce physical tension associated with excessive Beta activity, promoting relaxation while maintaining mental alertness.

## 2. Deep Breathing and Relaxation Exercises

Deep breathing exercises are an effective way to regulate Beta waves and maintain calmness under stress. Techniques such as **diaphragmatic breathing** or **4-7-8 breathing** can help lower stress-induced Beta waves by promoting relaxation in the body and mind.

- **Diaphragmatic Breathing:**

  By focusing on deep, diaphragmatic breaths, you activate the parasympathetic nervous system, which counterbalances the excessive Beta waves associated with stress. This helps bring the body and mind back into balance, allowing for greater focus and relaxation.

- **4-7-8 Breathing Technique:**

  Inhale for 4 seconds, hold the breath for 7 seconds, and exhale for 8 seconds. This technique promotes relaxation by regulating Beta activity and calming the nervous system, reducing the tendency toward mental overload.

## 3. Physical Exercise and Movement

Physical activity is one of the most effective ways to regulate Beta waves and reduce stress. Exercise helps reduce excessive Beta activity by releasing endorphins, which improve mood and provide a mental break from cognitive strain. Whether it's yoga, running, or strength training, exercise stimulates the brain to produce Beta waves in a balanced way, keeping you mentally alert and physically energized.

**Yoga and Tai Chi:**

## 4. Proper Sleep and Recovery

Sleep is critical for brainwave regulation. When you sleep, the brain cycles through various brainwave frequencies, including Delta and Theta waves. Sleep helps restore and refresh the mind, allowing for better Beta wave functioning during waking hours. Ensuring that you get enough rest and recovery time can help prevent Beta wave overload and maintain peak performance levels.

**Sleep Hygiene:**

## The Relationship Between Beta Waves and High Achievement

Beta waves are directly linked to high performance and achievement. Whether it's in sports, business, or any other field that requires focus, problem-solving, and decision-making, Beta waves are crucial for success. However, the key to achieving excellence is not just increasing Beta wave activity but learning how to manage it effectively. Striking the right balance allows individuals to harness the power of Beta waves to stay sharp, focused, and productive while avoiding the negative consequences of overstimulation.

Studies have shown that individuals who can regulate their Beta wave activity are more likely to excel in high-stakes environments, handle stress effectively, and maintain consistent performance over time. By using techniques like mindfulness, deep breathing, and regular physical exercise, individuals can maximize their Beta wave potential, leading to greater achievement and fulfillment in both their personal and professional lives.

## Conclusion

Beta waves are a critical component of peak performance. They enable quick thinking, concentration, and decision-making—skills that are essential for success in both sports and business. However, achieving optimal performance requires a delicate balance. By learning how to maintain high Beta wave activity without succumbing to stress or cognitive overload, individuals can excel in their chosen fields.

The techniques discussed in this chapter—mindfulness, relaxation, physical exercise, and proper sleep—provide practical tools for managing Beta wave activity and enhancing mental performance. In the following chapters, we will explore how Delta waves contribute to healing and sleep, and how Theta waves facilitate creativity and innovation. By mastering all brainwave frequencies, you can unlock your full potential and achieve high levels of performance and well-being.

# Chapter 14: Delta Waves for Healing and Deep Sleep

## The Science of Delta Waves in Sleep Cycles

Delta waves, which occur within the frequency range of 0.5 to 4 Hz, are the slowest brainwaves and are most commonly associated with deep sleep and profound restorative states. These waves dominate during the deepest stages of non-rapid eye movement (NREM) sleep, particularly in stages 3 and 4, which are essential for physical and mental recovery. While Delta waves are less commonly observed during waking hours, they are critical for maintaining long-term health and well-being.

During deep sleep, Delta waves facilitate critical physiological and neurological processes, including tissue repair, immune system strengthening, and memory consolidation. The body's restoration processes are optimized during Delta wave activity, making these brainwaves vital for health maintenance.

Research has shown that when Delta waves are prevalent during sleep, the brain becomes less responsive to external stimuli, enabling individuals to experience restorative, uninterrupted sleep. Additionally, Delta waves are linked to the body's ability to heal and regenerate, which is why deep sleep is often referred to as the "healing sleep."

## Delta Waves and the Sleep Cycle

Delta waves occur in the deepest stages of NREM sleep, known as slow-wave sleep. These stages are critical for the body's restorative processes and memory consolidation, as they facilitate:

- **Physical Healing:** Delta waves promote tissue repair and growth. During deep sleep, the body releases growth hormones that help regenerate tissues, muscles, and bones, contributing to recovery from physical exertion and injury.
- **Immune System Support:** Delta waves are involved in bolstering the immune system. Research has demonstrated that deep sleep boosts immune cell activity, helping the body fight infections and illnesses.
- **Cognitive Restoration and Memory Consolidation:** Delta waves also play a key role in consolidating memories from the day, transferring short-term memory into long-term storage. This process helps improve learning and problem-solving skills.

The quantity and quality of Delta sleep directly affect how well the body can recover, repair, and regenerate. An optimal amount of Delta sleep is crucial for maintaining both physical and mental health, as it restores the body's energy levels, enhances cognitive function, and promotes emotional well-being.

## Using Delta Waves to Promote Restful Sleep

Delta waves are crucial for achieving restorative sleep, and individuals can use various techniques to enhance Delta wave activity, ensuring deeper and more restorative sleep cycles. Leveraging Delta waves for sleep optimization can be especially beneficial for those suffering from sleep disorders or experiencing chronic sleep deprivation.

## 1. Sleep Optimization Techniques

- **Sleep Hygiene:**

  One of the most effective ways to promote Delta waves and improve sleep quality is to practice good sleep hygiene. Sleep hygiene includes creating a consistent bedtime routine, maintaining a cool and quiet sleep environment, and avoiding stimulants such as caffeine or electronic devices before bed. By cultivating a relaxed atmosphere conducive to sleep, you encourage the production of Delta waves, enabling you to enter deeper stages of sleep.

- **Pre-Sleep Relaxation Techniques:**

  Techniques such as progressive muscle relaxation (PMR) and deep breathing exercises can help prepare the body for restful sleep by reducing stress and facilitating the transition into deep sleep. These relaxation exercises encourage the onset of Delta waves, promoting deep, uninterrupted rest.

- **Mindfulness Meditation:**

  Practicing mindfulness meditation or focused breathing can also help activate Delta wave activity before sleep. Meditation helps reduce mental chatter and anxiety, creating a calm and relaxed state that facilitates Delta wave induction. This is particularly helpful for individuals who struggle with overactive minds before bedtime.

- **Listening to Delta Wave Stimuli:**

   Using binaural beats or isochronic tones designed to stimulate Delta wave activity is an effective way to encourage deeper sleep. Listening to these audio frequencies before bedtime can help synchronize your brainwaves, promoting a smoother transition into deep sleep and enhancing the overall quality of your sleep cycle.

## 2. Neurofeedback and Sleep Enhancement

Neurofeedback is a technique that helps individuals consciously regulate their brainwave activity. During a neurofeedback session, EEG sensors are placed on the scalp to monitor brainwaves, and feedback is provided to help individuals achieve specific brainwave patterns, such as increasing Delta wave activity.

For sleep optimization, neurofeedback sessions can be used to train individuals to enhance Delta waves during sleep. By using real-time feedback, individuals can learn to encourage Delta wave dominance during sleep cycles, resulting in deeper and more restorative sleep.

## How Delta Waves Aid in Tissue Repair and Cellular Regeneration

Delta waves not only contribute to a restful mind but also support the body's healing processes. This makes them critical for individuals recovering from physical exertion, injury, or illness. During the deep sleep phases characterized by Delta wave activity, the body releases human growth hormone (HGH), which is essential for tissue regeneration and cellular repair. As a result, Delta waves play a crucial role in:

- **Muscle Repair and Recovery:** After physical activity or muscle strain, Delta waves facilitate muscle regeneration by promoting HGH release, aiding in tissue repair and reducing soreness. This is why athletes and fitness enthusiasts often prioritize good sleep to ensure maximum recovery.

- **Cellular Regeneration and Immune System Boost:** Delta waves also play a role in cellular regeneration across the entire body, from skin cells to internal organs. Deep sleep, facilitated by Delta waves, allows the body to restore and regenerate cells, aiding in overall vitality and longevity.

- **Detoxification and Metabolic Health:** Sleep, especially Delta sleep, is when the brain and body detoxify, clearing away toxins and waste products that accumulate throughout the day. Adequate Delta wave activity ensures that this detoxification process occurs efficiently, contributing to overall health and energy levels.

## Delta Waves for Healing Emotional and Mental States

In addition to their physical healing properties, Delta waves also have an impact on mental and emotional health. Adequate Delta wave activity during sleep helps regulate emotions, reduce stress, and support emotional resilience. Here's how:

- **Emotional Recovery:** Delta waves help process and integrate emotional experiences from the day, ensuring that emotional wounds or stressors are resolved during deep sleep. This contributes to a balanced mood and emotional well-being.
- **Mental Health Benefits:** Studies have shown that improving Delta wave activity can help reduce symptoms of anxiety and depression. Delta sleep allows the brain to process and release emotional tension, leading to improved mental health and resilience.
- **Reducing Stress Hormones:** Delta waves are associated with the reduction of cortisol, a stress hormone. By enhancing Delta wave activity, individuals can lower cortisol levels, helping to relieve mental stress and promote a sense of peace and well-being.

## The Connection Between Delta Waves and Aging

As we age, the quality of deep sleep tends to decline, which can result in reduced Delta wave activity. This can have negative effects on physical health, cognitive function, and emotional well-being. Research has shown that older adults who experience deep sleep with sufficient Delta wave activity tend to have better memory, cognitive abilities, and immune function.

One of the primary reasons Delta wave activity decreases with age is due to changes in the brain's ability to regulate sleep cycles. However, promoting Delta waves through sleep optimization techniques, neurofeedback, and healthy lifestyle choices can help mitigate these effects and promote healthier aging.

## Conclusion

Delta waves are essential for healing, restorative sleep, and overall well-being. These brainwaves play a crucial role in physical recovery, immune system support, memory consolidation, and emotional regulation. By enhancing Delta wave activity through techniques like neurofeedback, mindfulness, and sound therapy, individuals can experience deeper sleep, faster recovery, and greater mental clarity.

For those struggling with sleep disorders or seeking to optimize their health, focusing on Delta wave enhancement can be a transformative tool. Delta waves offer profound healing benefits, not just for the body, but for the mind and emotions as well. By incorporating Delta wave-promoting practices into daily life, individuals can experience better sleep, improved physical health, and enhanced mental and emotional resilience.

In the following chapters, we will explore the role of Theta waves in creativity and problem-solving, as well as how Theta brainwave activity contributes to innovation and personal growth.

# Chapter 15: Harnessing Theta Waves for Creativity and Innovation

### Theta Waves and the Creative Process

Theta waves, oscillating at a frequency range of 4 to 8 Hz, are often associated with deep meditation, creativity, and the subconscious mind. These brainwaves are present during the transition between wakefulness and sleep and play a vital role in accessing deeper mental states. Theta waves are particularly potent in facilitating creativity because they allow for the free-flowing exchange of ideas, insights, and problem-solving strategies. They provide access to the subconscious, where memories, emotions, and intuitive knowledge reside.

When the brain enters a Theta state, it becomes less constrained by rational thought and logical analysis, allowing the imagination to flourish. This is why Theta waves are so powerful in unlocking creativity. The relaxed yet focused mental state associated with Theta waves encourages innovative thinking, artistic expression, and unique solutions to complex problems.

Theta waves are also deeply connected to the creative "Aha!" moments, where insights or solutions seem to appear out of nowhere. These moments often occur when the mind is in a relaxed state, and Theta waves help bridge the conscious and unconscious minds, facilitating access to ideas and concepts that would otherwise be hidden.

## The Role of Theta Waves in Problem-Solving

Theta waves play a significant role in problem-solving and innovative thinking. While Beta waves are linked to conscious thought and logical reasoning, Theta waves are associated with the ability to step back from conventional thinking and explore novel solutions. The subconscious mind, where Theta waves dominate, can make connections between seemingly unrelated ideas, leading to breakthroughs in creativity and innovation.

- **Nonlinear Thinking:**

  Theta waves encourage nonlinear thinking, which is essential for creativity and innovation. While Beta waves are effective for structured, analytical thinking, Theta waves help break free from traditional thought patterns, enabling the brain to consider different possibilities and unconventional solutions. This is why many creative professionals, such as artists, writers, and designers, find that they are most productive when they are relaxed and in a Theta-dominant mental state.

- **Enhanced Intuition and Insight:**

  As the brain enters the Theta state, intuitive insights become more accessible. The subconscious mind often provides guidance or answers to problems that the conscious mind struggles to solve. This intuitive knowledge is often experienced as "gut feelings" or sudden bursts of inspiration. The more one can tap into the Theta state, the more frequently these intuitive moments can occur.

- **Creative Visualization:**

  Theta waves are also linked to the practice of creative visualization. By visualizing success or outcomes in a relaxed, focused state, individuals can unlock creative potential and align their mental and emotional energy toward achieving specific goals. This visualization process helps in problem-solving by engaging the subconscious to create mental blueprints for success.

Techniques to Unlock Deep Creativity Using Theta Waves

There are several methods and techniques that can help increase Theta wave activity, which in turn unlocks creativity and fosters innovation. These approaches can be particularly useful for artists, writers, designers, or anyone looking to improve their creative output.

## 1. Meditation and Mindfulness Practices

Meditation is one of the most effective ways to enter a Theta-dominant state. Mindfulness meditation, specifically, encourages a state of relaxation and awareness that enables the brain to access Theta waves.

- **Guided Meditation:**

  Guided meditation can be particularly helpful for those new to accessing Theta states. By following a guide's voice, individuals can focus on specific visualizations or relaxations that promote Theta wave production. Many meditation apps offer guided sessions that target creative thinking, problem-solving, or intuition-building.

- **Open Monitoring Meditation:**

  Open monitoring meditation encourages awareness of all thoughts and sensations without attachment. This practice helps cultivate a deeper awareness of the present moment and facilitates the spontaneous emergence of creative ideas. By cultivating a relaxed, receptive state of mind, individuals can easily enter Theta states and allow their minds to freely explore new concepts.

## 2. Deep Breathing and Relaxation Techniques

Breathing techniques that promote deep relaxation can trigger the onset of Theta wave activity. Deep breathing helps slow the mind and body, facilitating the shift from Beta to Alpha, and then into the Theta state.

- **Diaphragmatic Breathing:**

  Diaphragmatic breathing, or belly breathing, focuses on slow, deep breaths that activate the parasympathetic nervous system. This process induces relaxation and lowers the levels of cortisol (the stress hormone), allowing the brain to shift toward a more creative and innovative mental state.

- **4-7-8 Breathing Technique:**

  Inhale for four seconds, hold for seven seconds, and exhale for eight seconds. This deep breathing technique helps relax the mind and body, facilitating Theta wave production and enhancing the ability to think creatively. It is especially useful before brainstorming sessions or moments where new ideas are needed.

## 3. Brainwave Entrainment

Brainwave entrainment through audio stimuli, such as binaural beats or isochronic tones, is a direct method to increase Theta wave activity. These sound frequencies help synchronize brainwave patterns with external stimuli, encouraging the brain to enter a Theta-dominant state.

- **Binaural Beats for Theta Waves:**

  Binaural beats are a form of brainwave entrainment that uses two slightly different frequencies in each ear to create a third frequency, perceived as a "beat" by the brain. Listening to Theta-inducing binaural beats can help the brain achieve deep relaxation and stimulate creative thought. These beats are widely available on meditation apps, YouTube, and specialized brainwave entrainment services.

- **Isochronic Tones:**

  Isochronic tones use single tones that pulse at specific intervals to induce brainwave synchronization. For creativity and innovation, isochronic tones that stimulate Theta waves can be highly effective. They offer a simple and effective way to access deep, relaxed states conducive to creative thinking and insight.

## 4. Visualization and Creative Thinking Exercises

Visualization is a powerful technique that helps harness the creative potential of Theta waves. Visualizing your goals, desired outcomes, or solutions to problems while in a relaxed state engages the subconscious mind and enhances creativity.

- **Visualization of Success:**

  Whether you are visualizing a successful project or imagining an ideal solution to a complex problem, creative visualization helps access Theta waves and unlock new perspectives. The more vividly and emotionally you engage with your visualization, the more easily the brain enters the Theta state.

- **Free Association and Brainstorming:**

  Another powerful tool for accessing Theta states is engaging in free association or brainstorming exercises. Allowing your mind to wander and connect seemingly unrelated ideas fosters the kind of creativity that Theta waves facilitate. The key is to remain open and non-judgmental about the ideas that surface, trusting that the Theta state will guide you to new and innovative solutions.

## How Theta Waves Facilitate Problem–Solving and Innovation

Theta waves are essential in the process of problem-solving and innovation. By promoting a relaxed yet focused mental state, Theta waves allow individuals to think outside the box and consider unconventional solutions. The Theta state encourages the brain to break free from rigid cognitive patterns, enabling new connections to form and innovative ideas to emerge.

- **Access to the Subconscious Mind:**

  The Theta state opens the door to the subconscious mind, where intuitive insights and solutions often reside. The brain can make connections between past experiences, emotions, and learned knowledge that would otherwise remain dormant. By accessing this wealth of subconscious information, individuals can generate creative ideas, solve complex problems, and innovate.

- **Enhanced Pattern Recognition:**

  Theta waves enhance the brain's ability to recognize patterns and see relationships between disparate pieces of information. This is crucial for solving complex problems and discovering innovative solutions. The Theta state enables the mind to "see the bigger picture," connecting dots that might otherwise be missed.

## Conclusion

Theta waves play a pivotal role in fostering creativity, innovation, and problem-solving. By accessing the Theta state, individuals can unlock their full creative potential, tap into their subconscious mind, and approach challenges with fresh perspectives. Through techniques like meditation, deep breathing, brainwave entrainment, and creative visualization, you can increase Theta wave activity and enhance your ability to innovate.

In the following chapters, we will delve into practical tools for brainwave training and explore how mastering Theta waves can lead to personal growth and transformative change. By harnessing the power of Theta waves, you can cultivate creativity, solve problems more effectively, and unleash your full potential.

# Chapter 16: Brainwave Training: Methods and Tools

## Practical Tools for Brainwave Training

Mastering neural oscillations—whether it's enhancing the focus associated with Beta waves, boosting creativity through Theta waves, or facilitating deep relaxation via Alpha waves—requires both understanding and consistent practice. Brainwave training empowers individuals to harness these states at will, improving cognitive performance, emotional regulation, and overall well-being. There are several effective tools and methods available for training brainwave activity, ranging from traditional techniques like meditation to cutting-edge technologies such as neurofeedback.

This chapter will delve into the various methods and tools available for brainwave training, highlighting practical strategies that can help you gain control over your neural oscillations.

## 1. Meditation and Mindfulness Techniques

Meditation is one of the most powerful and accessible methods for training and controlling brainwaves. While all types of meditation encourage relaxation and stress relief, specific techniques can stimulate certain brainwave frequencies. With consistent practice, meditation allows you to enter desired brainwave states, such as Alpha for relaxation or Theta for deep creativity.

## Mindfulness Meditation for Alpha and Theta Waves:

Mindfulness meditation involves paying attention to the present moment without judgment. By focusing on the breath or bodily sensations, you shift from a busy, Beta-dominant mind to a more relaxed Alpha state. This process can also encourage light Theta activity, promoting creativity and emotional clarity. Practicing mindfulness consistently can increase your ability to enter these brainwave states whenever needed.

### Technique:

## Focused Attention Meditation for Beta Waves:

On the other hand, Beta waves, associated with concentration and focus, can be trained through focused attention meditation. This practice involves concentrating on a single object or thought without distraction. While this meditation style encourages mental clarity and alertness, it can also train the brain to maintain healthy Beta wave activity for peak cognitive performance.

**Technique:**

## 2. Neurofeedback: The Cutting-Edge Brainwave Training

Neurofeedback is an advanced tool that provides real-time feedback on your brainwave patterns. This technique uses an electroencephalogram (EEG) to monitor brainwave activity and gives immediate feedback, allowing individuals to learn how to control their brainwaves. Neurofeedback has become a popular method for training various brainwave frequencies, including Alpha, Beta, and Theta waves.

### How Neurofeedback Works:

Neurofeedback involves placing electrodes on the scalp to measure brainwave activity. The brain's electrical activity is then visualized on a computer screen in real-time. When the brain produces the desired brainwave patterns, individuals are rewarded with positive feedback, such as a visual cue or sound. Over time, the brain learns to produce these desired brainwave states more naturally.

- **Applications in Alpha Waves:**

  Neurofeedback can be used to increase Alpha wave activity, which is associated with relaxation, stress relief, and creativity. By using neurofeedback to reward Alpha-dominant brainwaves, you can train your brain to enter this relaxed state more easily, leading to improved emotional regulation and cognitive function.

- **Applications in Theta Waves:**

  Theta waves, linked to creativity and deep meditation, can also be enhanced through neurofeedback. This type of training encourages the brain to enter a state that fosters insight, relaxation, and enhanced intuitive thinking. Many creative professionals use neurofeedback to access deeper levels of innovation and creative flow.

## Benefits of Neurofeedback:

- **Improved Emotional Regulation:**

  Neurofeedback can help you regulate brainwave patterns, leading to better control over your emotional responses.

- **Enhanced Cognitive Function:**

  By training Beta waves, neurofeedback can also improve focus, memory, and problem-solving skills.

- **Customized Brainwave Training:**

  Neurofeedback can be tailored to meet your specific needs, whether you want to reduce anxiety, increase creativity, or optimize cognitive performance.

### 3. Brainwave Entrainment: Harnessing Sound and Light

Brainwave entrainment is the process of using external stimuli—such as sound or light—to synchronize brainwave activity to a specific frequency. One of the most popular forms of brainwave entrainment is **binaural beats**.

### Binaural Beats for Brainwave Training:

Binaural beats work by playing two slightly different frequencies in each ear. The brain perceives a third "beat" as the difference between the two frequencies, which leads to the brain adopting a frequency corresponding to that beat. For example, if one ear hears a 300 Hz tone and the other hears a 310 Hz tone, the brain will perceive a beat of 10 Hz, which corresponds to the Alpha brainwave range.

- **How It Works:**
  To achieve a desired brainwave state, you can listen to audio tracks specifically designed to target the frequency of the brainwave you want to enhance. For instance, binaural beats designed for relaxation may use a 10 Hz beat (Alpha range), while binaural beats designed to boost concentration could use a frequency closer to 20 Hz (Beta range).

- **Applications in Theta Waves:**
  For deep creativity or meditation, Theta-inducing binaural beats can help guide the brain into the desired state. By listening to Theta-focused binaural beats, individuals can promote relaxed, intuitive thinking, and creativity.

## Isochronic Tones for Precision Entrainment:

Unlike binaural beats, which require headphones, isochronic tones use single, pulsating sounds to entrain brainwaves. The rhythm of the sound pulses at a specific frequency, directly influencing brainwave activity. Isochronic tones are particularly effective for deep meditation or increasing focus.

## How It Works:

## Benefits of Brainwave Entrainment:

- **Improved Focus and Cognitive Performance:**

  Listening to Beta-inducing tracks can improve concentration and problem-solving skills, making it an excellent tool for students and professionals.

- **Stress Reduction and Relaxation:**

  Alpha-inducing sounds can help reduce stress, anxiety, and mental fatigue, providing an effective tool for relaxation.

- **Enhanced Creativity:**

  Theta-inducing tracks facilitate creative thinking, insight, and access to the subconscious, making them an excellent tool for creative professionals.

## 4. Biofeedback: Measuring and Modifying Brainwave Activity

Biofeedback is a technique that allows individuals to monitor and control their body's physiological functions, such as heart rate, muscle tension, and brainwave activity. Through biofeedback, individuals receive immediate feedback on their brainwaves and can use this information to consciously modify their brainwave patterns.

## How Biofeedback Works:

- Sensors are attached to the scalp, providing real-time data on brainwave activity.
- The individual is trained to control their brainwave patterns using visual or auditory feedback.
- Over time, this feedback helps individuals learn how to produce specific brainwave states, such as increasing Alpha waves for relaxation or Theta waves for creativity.

Biofeedback is effective for training both conscious and subconscious control over brainwave activity, and it can be used in conjunction with meditation, neurofeedback, or other techniques to enhance overall brain function.

## 5. Measuring Progress and Refining Brainwave Control

Tracking your progress is essential to mastering brainwave control. Whether using neurofeedback, biofeedback, or other methods, measuring your brainwave activity can help refine your techniques and improve the consistency of your results. Tools like EEG devices, wearable brainwave monitors, and biofeedback systems provide real-time data on your brainwave patterns, allowing you to adjust your training methods as needed.

## Monitoring Tools for Brainwave Training:

- **EEG Devices:** These devices provide direct measurement of brainwave activity and are commonly used in neurofeedback training.

- **Wearable Brainwave Monitors:** Portable EEG devices can track brainwave activity throughout the day, allowing users to measure progress and make adjustments to their techniques.

- **Biofeedback Devices:** These tools allow for real-time monitoring of physiological data, including brainwave activity, and help users fine-tune their training for optimal results.

## Conclusion

Brainwave training offers a wide range of tools and techniques to help you harness the power of your brain's natural rhythms. Whether through meditation, neurofeedback, brainwave entrainment, or biofeedback, there are multiple methods for gaining control over your neural oscillations. By mastering your brainwaves, you can enhance your cognitive abilities, emotional intelligence, creativity, and overall well-being.

In the following chapters, we will explore how mastering brainwaves can lead to personal growth, improved physical performance, and long-term success. By incorporating these brainwave training techniques into your daily routine, you can unlock your full potential and achieve a higher level of mental, emotional, and physical well-being.

# Chapter 17: Applications in Personal Growth

### How Mastering Brainwaves Can Enhance Self-Awareness and Emotional Intelligence

Mastering brainwaves provides a direct pathway to enhancing various aspects of personal growth, including self-awareness, emotional intelligence, creativity, and mental resilience. By understanding and controlling brainwave activity, individuals can fine-tune their mental states, enabling them to better manage their emotions, thoughts, and behaviors.

Self-awareness is the foundation of personal growth. It involves the ability to observe one's own thoughts and emotions without judgment. By consciously controlling brainwave patterns, such as Alpha and Theta waves, individuals can access deeper layers of their consciousness, uncover hidden aspects of themselves, and cultivate greater self-awareness. This enhanced awareness provides clarity, allowing for more effective decision-making, increased confidence, and improved interpersonal relationships.

Moreover, emotional intelligence (EQ) is another key area of personal growth that is greatly influenced by brainwave mastery. Emotional intelligence involves the ability to recognize, understand, and manage one's own emotions, as well as the ability to empathize with others. By promoting Alpha waves for emotional regulation or Theta waves for intuition, individuals can better respond to emotional triggers, reduce stress, and develop more empathetic and constructive interactions.

## Key Brainwave States for Personal Growth

- **Alpha Waves (8–12 Hz):**

  Alpha waves are associated with relaxation and flow. Cultivating Alpha wave activity can help individuals manage stress and anxiety, as well as access a state of calm that allows for clearer thinking and emotional balance. By incorporating techniques to increase Alpha waves, such as mindfulness meditation, individuals can reduce mental clutter and become more attuned to their emotional states.

- **Theta Waves (4–8 Hz):**

  Theta waves are closely connected to deep meditation, creativity, and access to the subconscious mind. By training the brain to produce more Theta waves, individuals can tap into their creative potential, gain insights into their deepest desires and motivations, and enhance their problem-solving abilities. Theta wave activity also plays a key role in memory consolidation, which can aid in emotional healing and self-discovery.

- **Beta Waves (12–30 Hz):**

  Beta waves are essential for focused attention, logical thinking, and decision-making. While excessive Beta activity can lead to stress and overthinking, controlled Beta waves can help individuals stay mentally alert and focused during challenging tasks. Finding a healthy balance of Beta activity can help individuals maintain high performance without burnout.

Brainwave-Based Techniques for Personal Transformation

Mastering brainwaves provides a range of techniques that can be used for personal transformation. These techniques can help you optimize your cognitive abilities, enhance your emotional intelligence, and facilitate long-term growth.

## 1. Neurofeedback for Emotional Regulation

Neurofeedback is one of the most effective tools for brainwave mastery. By using real-time brainwave data, neurofeedback enables individuals to train their brains to produce specific brainwave patterns. For emotional regulation, neurofeedback can help individuals increase Alpha wave activity, facilitating relaxation and emotional stability. This technique has been used in clinical settings to treat anxiety, depression, and PTSD, with individuals reporting increased emotional resilience and self-awareness.

## 2. Meditation for Self-Awareness and Mindfulness

Meditation is a powerful method for tuning into one's inner world. By using specific meditation techniques, individuals can increase their ability to enter Theta or Alpha states, unlocking deeper levels of self-awareness and mindfulness. For example, loving-kindness meditation (Metta) can promote empathy and emotional intelligence by training individuals to cultivate feelings of compassion for themselves and others. Through consistent practice, meditation can transform one's thought patterns, helping to foster greater emotional balance and insight.

- **Mindfulness Meditation:**

  Mindfulness meditation encourages present-moment awareness and non-judgmental observation of thoughts, emotions, and bodily sensations. This practice helps reduce mental noise and allows for increased self-awareness. It also trains individuals to become more attuned to their emotional responses, promoting emotional regulation.

- **Guided Visualization:**

  Guided visualization, often used to enhance Theta waves, involves imagining specific outcomes or exploring deep-seated emotions. By visualizing desired goals, individuals can engage their subconscious mind to manifest positive changes in their lives. Visualization can also be used to heal past emotional wounds by revisiting memories and re-framing them in a healing light.

### 3. Journaling and Brainwave Synchronization

Journaling is an excellent tool for promoting self-awareness and emotional intelligence. Writing down thoughts and emotions allows individuals to process and reflect on their experiences. By incorporating brainwave entrainment (such as listening to binaural beats or isochronic tones) while journaling, individuals can enhance the connection between their conscious mind and subconscious processes.

This synchronization of brainwaves can lead to more profound insights, emotional clarity, and a deeper understanding of personal desires and motivations. Combining journaling with brainwave entrainment can also amplify creativity and problem-solving abilities, providing a holistic approach to personal growth.

### 4. Biofeedback for Emotional Intelligence Development

Biofeedback is another technique that can enhance emotional intelligence by helping individuals become aware of their physiological responses to stress and emotional triggers. With the help of biofeedback devices that monitor heart rate variability (HRV), skin conductivity, and brainwave activity, individuals can learn to regulate their autonomic nervous system and brainwave patterns to respond more effectively to emotional challenges.

By training to increase coherence in heart rate and brainwaves, individuals can improve their ability to remain calm and focused during stressful situations, ultimately enhancing their emotional intelligence.

### Case Studies of Individuals Who Have Enhanced Their Lives Through Brainwave Mastery
### Case Study 1: Achieving Emotional Balance with Alpha Waves

Jennifer, a corporate executive, had been struggling with chronic stress and anxiety. After engaging in neurofeedback sessions designed to increase Alpha wave activity, Jennifer began to experience significant improvements in her emotional well-being. The neurofeedback helped her learn to produce more Alpha waves, which in turn allowed her to manage stress more effectively. She found that she was more patient with her colleagues, less reactive to stressful situations, and had a greater sense of inner peace.

## Case Study 2: Unlocking Creativity with Theta Waves

David, an aspiring writer, had been experiencing creative blocks that hindered his progress on his novel. After committing to daily meditation sessions focused on increasing Theta wave activity, David discovered an incredible surge in his creativity. By accessing the Theta state, he tapped into his subconscious mind, allowing his ideas to flow freely and effortlessly. Within weeks, he completed several chapters of his book, feeling more inspired than ever.

## Case Study 3: Improving Focus and Productivity with Beta Waves

Sarah, a university student, struggled with maintaining focus during her studies, often feeling distracted and overwhelmed. After incorporating Beta wave training through brainwave entrainment (listening to binaural beats while studying), Sarah noticed a significant improvement in her concentration. The increase in Beta waves allowed her to maintain mental clarity and focus for longer periods, improving her academic performance and productivity.

## Conclusion

Mastering brainwaves is a transformative tool for personal growth. By consciously training specific brainwave states such as Alpha, Beta, and Theta, individuals can enhance their emotional intelligence, self-awareness, and creativity. Whether through meditation, neurofeedback, biofeedback, or brainwave entrainment, these techniques provide practical pathways to achieving greater mental clarity, emotional resilience, and overall well-being.

The power to shape your mind is in your hands. By incorporating brainwave mastery into your daily life, you can unlock your full potential, create lasting change, and enhance the quality of your life in profound ways.

# Chapter 18: Brainwaves and Physical Performance

## The Impact of Brainwave Regulation on Physical Training

When most people think of physical performance, they often focus on the body's strength, stamina, or endurance. However, the brain plays a crucial role in physical performance as well. Our brainwaves directly influence how we move, how we control our muscles, and how effectively we respond to physical challenges. By regulating brainwave activity, individuals can optimize their physical performance, improve recovery times, and even enhance their overall physical health.

The connection between brainwave regulation and physical performance can be understood by recognizing that different brainwave states influence various aspects of motor control, focus, relaxation, and recovery. For example, Beta waves, associated with concentration, can help enhance coordination and focus during physical activities, while Delta waves, associated with deep sleep and recovery, play a key role in muscle repair and overall bodily restoration.

In this chapter, we'll explore how different brainwave frequencies—Alpha, Beta, Delta, and Theta—affect physical training and performance, and how mastering these brainwaves can improve athletic performance, physical recovery, and overall well-being.

## Using Alpha Waves for Relaxation and Coordination

Alpha waves, ranging from 8 to 12 Hz, are associated with a state of calm relaxation, focus, and mental clarity. While Alpha waves are often linked to stress reduction and mental relaxation, they also have a direct impact on physical performance.

**Role of Alpha Waves in Physical Performance:**

- **Enhancing Coordination:**

  During physical activities, Alpha waves can help improve hand-eye coordination and fine motor skills. Athletes who enter an Alpha state while performing a task—whether it's playing a sport, practicing a musical instrument, or even lifting weights—often experience greater fluidity and accuracy in their movements.

- **Promoting Relaxed Focus:**

  Many athletes experience peak performance when they are in a relaxed but focused state, often referred to as "flow." Alpha waves support this relaxed concentration by balancing the mind's need for focus with the body's need for relaxation. This is particularly important in activities such as yoga, Pilates, or sports like tennis or golf, where mental relaxation and physical precision are crucial.

- **Reducing Muscle Tension:**

  Alpha waves have the ability to relax the body's muscle tension, making them an excellent tool for athletes who need to reduce tightness and improve their range of motion. Practices like progressive muscle relaxation and mindful breathing, both of which promote Alpha wave production, are effective in releasing physical tension and improving flexibility.

## Techniques for Enhancing Alpha Waves:

- **Mindfulness Meditation:**

  Practicing mindfulness meditation, focusing on the breath and being present in the moment, can enhance Alpha wave activity, promoting relaxation and mental clarity during physical activities.

- **Breathing Techniques:**

  Slow, deep breathing exercises, such as diaphragmatic breathing, activate the parasympathetic nervous system and increase Alpha wave production, leading to a more relaxed state that can be beneficial in physical performance.

## Using Beta Waves for Focus and Alertness in Performance

Beta waves, ranging from 12 to 30 Hz, are associated with heightened alertness, concentration, and mental activity. These brainwaves play an essential role in tasks that require focus, attention, and mental effort, such as decision-making or solving problems in real-time.

## Role of Beta Waves in Physical Performance:

- **Mental Alertness in High-Stress Situations:**

  Beta waves help athletes stay mentally alert and focused during high-pressure situations. Whether it's a football quarterback making a split-second decision, a sprinter preparing for a race, or a gymnast performing a routine, Beta waves allow for quick thinking and fast reflexes, which are essential in fast-paced sports and high-stakes environments.

- **Maximizing Physical Performance:**

  Beta wave activity is necessary when the body is pushing to its physical limits, such as during weightlifting or sprinting. The heightened focus and mental engagement provided by Beta waves allow athletes to push through physical barriers, improving strength, stamina, and endurance.

- **Improved Reaction Times:**

  Beta waves play a crucial role in enhancing reaction time, which is essential for athletes in sports that require quick reflexes, such as boxing, tennis, or basketball. Increased Beta activity helps the brain process visual and sensory information more quickly, allowing athletes to react faster to external stimuli.

## Techniques for Enhancing Beta Waves:

- **Concentration and Focus Drills:**

  Engaging in exercises that require sustained attention, such as visual tracking exercises or mental puzzles, can help improve Beta wave activity. These exercises simulate the mental effort needed during a physical performance and can enhance an athlete's ability to stay mentally alert and focused.

- **Brainwave Entrainment:**

  Listening to audio tracks designed to stimulate Beta waves through binaural beats or isochronic tones can help athletes increase their focus and alertness in preparation for competitions or training.

## Leveraging Delta Waves for Recovery and Healing

Delta waves, which range from 0.5 to 4 Hz, are the slowest brainwaves and are primarily associated with deep sleep and physical restoration. These waves are critical for healing processes, muscle repair, and overall physical rejuvenation.

## Role of Delta Waves in Physical Recovery:

- **Promoting Deep Sleep and Muscle Repair:**

  Delta waves are dominant during deep, restorative sleep, the phase when the body undergoes the most significant physical repair. Muscle tissue repairs, hormones are released for growth and recovery, and immune function is optimized during the Delta state. Adequate Delta wave activity during sleep is essential for athletes who need to recover after intense training or competition.

- **Restoring Energy Levels:**

  Delta waves help the body enter a restorative state, reducing fatigue and restoring energy. This is vital for athletes who engage in strenuous physical activity that depletes energy stores. Delta wave activity accelerates the recovery process, helping athletes feel rejuvenated and ready for their next training session.

- **Reducing Inflammation and Enhancing Tissue Healing:**

  Delta waves are linked to tissue regeneration and the reduction of inflammation. Regular, quality sleep that promotes Delta wave production can help reduce the risk of injury and speed up the healing process if an injury occurs.

## Techniques for Enhancing Delta Waves:

- **Sleep Optimization:**

  Ensuring that you get enough deep sleep each night is key to enhancing Delta wave activity. Creating a relaxing sleep environment, avoiding stimulants before bed, and following a consistent sleep schedule can improve the quality and duration of Delta wave-rich deep sleep.

- **Relaxation and Visualization:**

  Using relaxation techniques such as progressive muscle relaxation or guided visualization before sleep can encourage the onset of Delta waves, enhancing physical recovery during sleep.

## Using Theta Waves for Muscle Awareness and Mind-Body Connection

Theta waves, which oscillate between 4 and 8 Hz, are often linked to creativity and deep relaxation. However, Theta waves can also enhance the mind-body connection, making them valuable during physical activities that require body awareness and coordination.

## Role of Theta Waves in Physical Performance:

- **Enhanced Body Awareness:**

  Theta waves are associated with deep relaxation and body awareness, allowing athletes to tune into their bodies during complex movements. This mind-body connection can enhance an athlete's ability to perform activities like yoga, pilates, or martial arts, where precise movements and balance are essential.

- **Improved Flow States in Physical Activities:**

  Theta waves are central to the flow state, a mental state where athletes experience effortless performance. Theta-induced flow enhances performance by minimizing mental distractions, allowing athletes to perform at their peak without overthinking or forcing movements.

## Techniques for Enhancing Theta Waves:

- **Mindfulness and Meditation Practices:**

  Deep meditation and mindfulness practices, such as body scan meditation or mindful walking, can enhance Theta wave activity, improving body awareness and promoting a sense of calm and focus during physical activities.

- **Creative Visualization for Performance:**

  Visualizing athletic performance in a Theta-dominant state can also improve coordination, creativity, and mental preparedness. By envisioning yourself executing movements with fluidity and precision, you engage Theta waves that enhance performance during actual practice or competition.

## Conclusion

Brainwave regulation plays a vital role in optimizing physical performance and recovery. Alpha waves enhance coordination, relaxation, and focus; Beta waves boost alertness, concentration, and reaction time; Delta waves accelerate recovery and muscle repair; and Theta waves foster body awareness and flow. By understanding and mastering these brainwave states, athletes can improve their performance, reduce the risk of injury, and recover more effectively.

Incorporating brainwave training into your physical routine—through meditation, neurofeedback, sleep optimization, and other techniques—can help you reach peak physical health and performance. Whether you are an elite athlete or someone who wants to improve your fitness levels, brainwave mastery is a key component of physical success.

# Chapter 19: The Impact of Diet and Lifestyle on Brainwaves

## Nutritional Factors That Influence Brainwave Patterns

The food we eat and the lifestyle choices we make directly impact the functioning of our brain, including the generation and regulation of brainwaves. Certain nutrients and dietary practices can help optimize brainwave activity, improving cognitive function, emotional regulation, and physical health. Conversely, poor dietary choices and lifestyle habits can create imbalances in brainwave patterns, potentially leading to issues like anxiety, depression, fatigue, and cognitive decline.

In this chapter, we'll explore the connection between diet, lifestyle, and brainwave activity. By understanding how nutritional factors influence brainwaves, you can make informed choices to promote optimal mental and physical health, enhance cognitive performance, and maintain emotional stability.

## The Role of Macronutrients in Brainwave Activity

Macronutrients—proteins, fats, and carbohydrates—serve as the building blocks for brain health. They provide the energy and materials required for brain cells to function properly, which directly affects brainwave generation.

## 1. Proteins and Amino Acids:

Proteins are essential for the production of neurotransmitters, the chemical messengers in the brain that regulate brainwave activity. Amino acids, the building blocks of proteins, play a critical role in the synthesis of these neurotransmitters. For example, serotonin, dopamine, and GABA are neurotransmitters that influence mood, focus, and relaxation, respectively.

- **Serotonin**: This neurotransmitter is associated with feelings of well-being and happiness and plays a role in Alpha wave generation. Consuming foods rich in tryptophan (an amino acid precursor to serotonin) like turkey, nuts, seeds, and tofu can enhance Alpha wave activity, which is beneficial for relaxation and stress reduction.
- **Dopamine**: Known as the "reward" neurotransmitter, dopamine is linked to motivation, pleasure, and concentration. Higher dopamine levels promote Beta wave activity, which is essential for focus, alertness, and problem-solving. Foods like lean meats, fish, eggs, and dairy products provide tyrosine, an amino acid necessary for dopamine production.
- **GABA (Gamma-Aminobutyric Acid)**: GABA helps inhibit excessive neural firing and promotes relaxation by increasing Alpha wave activity. GABA-rich foods like spinach, broccoli, and almonds support relaxation and can reduce stress and anxiety.

## 2. Fats and Omega-3 Fatty Acids:

Healthy fats, particularly omega-3 fatty acids, play a crucial role in brain health. Omega-3s are vital for the structure and function of neurons and help regulate brainwave activity.

- **Omega-3 Fatty Acids**: Found in fatty fish (like salmon, mackerel, and sardines), flaxseeds, walnuts, and chia seeds, omega-3s are essential for cognitive function and mental clarity. Regular consumption of omega-3s has been linked to increased Alpha wave activity, improving focus and relaxation.
- **Monounsaturated Fats**: These fats, found in olive oil, avocados, and nuts, support brain health by enhancing neuronal communication. They help modulate brainwave activity and support cognitive function by improving memory and focus.

### 3. Carbohydrates and Brainwave Regulation:

Carbohydrates are a primary source of energy for the brain, and they influence brainwave patterns by regulating blood sugar levels. However, the type of carbohydrates you consume can affect brainwave activity in different ways.

- **Complex Carbohydrates**: Whole grains, legumes, and vegetables are rich in complex carbohydrates, which provide a steady release of glucose to the brain, supporting stable brainwave activity. These foods can enhance mental clarity and promote relaxation by stabilizing blood sugar levels.

- **Simple Carbohydrates**: On the other hand, refined sugars and processed foods that contain simple carbohydrates can cause blood sugar spikes followed by crashes. This can lead to imbalances in brainwave activity, often resulting in periods of low energy, irritability, and difficulty concentrating.

## Micronutrients and Their Impact on Brainwaves

Micronutrients, including vitamins and minerals, are essential for proper brain function and regulation of brainwave activity. Many of these nutrients are involved in the production of neurotransmitters and other compounds that influence cognitive and emotional health.

### 1. B Vitamins:

B vitamins, including B6, B12, and folic acid, are vital for brain function. They support the synthesis of neurotransmitters and are crucial for energy metabolism.

- **Vitamin B6**: This vitamin is involved in the production of GABA, which promotes relaxation and enhances Alpha wave activity. Foods rich in B6, such as poultry, fish, and bananas, help support calmness and emotional balance.
- **Vitamin B12 and Folate**: B12 and folate are essential for maintaining healthy brain cells and preventing cognitive decline. Deficiencies in these vitamins can lead to poor concentration and brain fog. Rich sources of B12 include meat, dairy, and fortified cereals, while folate is found in leafy greens, beans, and citrus fruits.

## 2. Magnesium:

Magnesium is a crucial mineral for brain function and the regulation of brainwave activity. It has a calming effect on the nervous system and helps promote Alpha and Theta wave activity.

### Magnesium-Rich Foods

## 3. Antioxidants and Brain Health:

Antioxidants, such as vitamins C and E, protect the brain from oxidative stress and inflammation, which can impair cognitive function and disrupt brainwave patterns.

### Blueberries and Dark Chocolate

## Lifestyle Choices That Influence Brainwave Activity

While diet plays a crucial role in brainwave regulation, lifestyle choices such as sleep, physical activity, and stress management are equally important in maintaining optimal brainwave patterns.

## 1. Sleep:

Sleep is one of the most important factors influencing brainwave activity. During deep sleep (dominated by Delta waves), the brain undergoes vital restorative processes, such as memory consolidation, detoxification, and tissue repair.

**Sleep Hygiene**

## 2. Physical Exercise:

Regular physical activity has been shown to increase the production of beneficial brain chemicals like endorphins and BDNF (brain-derived neurotrophic factor). These substances enhance cognitive function, promote brain plasticity, and improve brainwave regulation.

- **Aerobic Exercise**: Activities like running, swimming, or cycling can increase the production of Alpha and Theta waves, helping to improve mental clarity and reduce stress.
- **Strength Training**: Weightlifting and resistance training can boost Beta wave activity, enhancing focus and mental alertness during physical tasks.

## 3. Stress Management:

Chronic stress can lead to an imbalance in brainwave activity, particularly increasing Beta waves to unhealthy levels. This heightened Beta activity can cause anxiety, restlessness, and burnout.

**Mindfulness Practices**

### 4. Social Interaction and Mental Engagement:

Social interaction and mental stimulation help keep the brain active and engaged. Activities like socializing, problem-solving, or engaging in creative pursuits increase brainwave coherence, particularly in the Alpha and Theta ranges.

**Engaging Hobbies**

### Creating a Lifestyle That Supports Optimal Brainwave Function

Achieving optimal brainwave activity requires a holistic approach that includes a balanced diet, regular exercise, sufficient sleep, and effective stress management. By making informed choices about what we eat, how we move, and how we manage our mental health, we can create a lifestyle that supports our brain's natural rhythms and enhances cognitive performance, emotional well-being, and physical health.

Incorporating brainwave-enhancing foods and practices into your daily life will help regulate your brainwaves, promote mental clarity, reduce stress, and improve overall well-being. Whether you're looking to boost your productivity, enhance your creativity, or simply feel more at ease, optimizing your brainwave activity through diet and lifestyle choices is a powerful tool for unlocking your full potential.

# Chapter 20: Overcoming Brainwave Imbalances

### Identifying Brainwave Imbalances and Their Symptoms

Brainwaves are the electrical signals generated by the brain, and their specific patterns are crucial for different mental states and cognitive functions. However, when these brainwave patterns become imbalanced—whether too high or too low—it can result in various cognitive, emotional, and behavioral issues. Recognizing these imbalances is the first step in restoring optimal brainwave activity.

Imbalances in brainwaves typically manifest in several ways, including difficulty focusing, poor memory, anxiety, irritability, or even more severe issues like depression and insomnia. In this chapter, we will explore common brainwave imbalances, their symptoms, and how they can affect mental and physical health. We will also discuss strategies and techniques for restoring balance and optimizing brainwave function through different practices.

# Common Brainwave Imbalances

Brainwave imbalances occur when the frequency of brainwave activity in specific regions of the brain is not in alignment with normal, healthy patterns. These imbalances can be caused by a variety of factors, including stress, lack of sleep, poor nutrition, or neurological conditions. Below, we discuss the most common types of brainwave imbalances and how they affect cognitive and emotional health.

## 1. Alpha Wave Imbalance

Alpha waves (8–12 Hz) are associated with states of calmness, relaxation, and creativity. Imbalances in Alpha wave activity can manifest in the following ways:

- **Low Alpha Wave Activity**: Insufficient Alpha wave production is often linked to stress, anxiety, and hyperactivity. It can cause difficulty in relaxing, increased mental chatter, and trouble focusing. People with low Alpha wave activity often experience restlessness and are easily distracted. This can affect creativity, problem-solving abilities, and overall emotional balance.

- **High Alpha Wave Activity**: While Alpha waves are beneficial for relaxation, an excess of them can lead to sluggishness, lack of motivation, and disengagement. This can result in a person feeling lethargic or too passive, making it difficult to perform high-concentration tasks.

## 2. Beta Wave Imbalance

Beta waves (12–30 Hz) are typically associated with focused attention, alertness, and active thinking. However, imbalances in Beta wave activity can lead to cognitive and emotional difficulties:

- **High Beta Wave Activity**: Excessive Beta waves, especially in the higher frequencies (above 20 Hz), are often related to stress, anxiety, and even panic. People with high Beta wave activity may experience constant mental tension, overthinking, and difficulty relaxing. They are also more prone to mental fatigue, burnout, and stress-related disorders.

- **Low Beta Wave Activity**: Insufficient Beta waves can lead to lack of focus, poor concentration, and problems with processing information quickly. Individuals with low Beta activity might struggle with staying alert, have trouble organizing their thoughts, and show a general lack of motivation.

### 3. Theta Wave Imbalance

Theta waves (4–8 Hz) are primarily associated with deep relaxation, creativity, and access to the subconscious mind. Theta imbalances can cause a variety of issues:

- **Low Theta Wave Activity**: Insufficient Theta wave production is often linked to an inability to access deep creativity or intuitive thinking. People with low Theta activity may feel mentally blocked, lack imagination, or struggle with innovative problem-solving. They might also experience difficulty relaxing or meditating deeply.

- **High Theta Wave Activity**: An overabundance of Theta waves can lead to excessive daydreaming, a lack of focus, and an inability to stay grounded in the present moment. While Theta waves are beneficial for creativity, too much can result in a disconnected, unproductive state. People with high Theta activity may struggle with attention deficit or dissociation.

## 4. Delta Wave Imbalance

Delta waves (0.5–4 Hz) are the slowest brainwaves and are most prominent during deep sleep. Imbalances in Delta waves can significantly impact restorative processes and overall health:

- **Low Delta Wave Activity**: Low Delta waves are often associated with poor sleep quality, particularly with individuals who suffer from insomnia or other sleep disorders. A lack of Delta wave activity during sleep can prevent the brain from entering deep restorative states, leading to sleep deprivation, impaired healing, and cognitive decline.

- **High Delta Wave Activity**: Excessive Delta waves during waking hours can lead to feelings of deep fatigue, lack of motivation, and difficulty concentrating. This can make individuals feel disconnected or sluggish during daily activities and impair their cognitive abilities.

## Techniques for Restoring Balance Between Brainwave States

Restoring balance to brainwave activity is essential for maintaining cognitive and emotional health. There are several methods that can help regulate brainwave patterns and promote optimal brain function.

### 1. Meditation and Mindfulness Practices

Meditation is one of the most effective tools for regulating brainwave activity. Different forms of meditation can influence specific brainwave states, helping to restore balance in cases of brainwave imbalances.

- **For Alpha Wave Balance**: Meditation techniques that focus on deep relaxation, such as mindfulness meditation, can help increase Alpha wave activity and reduce stress and anxiety. Guided imagery or breathing exercises can promote a calm and focused state.

- **For Beta Wave Balance**: To manage high Beta wave activity, mindfulness practices such as focused attention meditation can help reduce mental tension and overthinking. Deep breathing exercises, progressive muscle relaxation, and body scan meditations can activate the parasympathetic nervous system, helping to calm the mind and body.

- **For Theta Wave Balance**: Practices like deep meditation or guided visualization can help boost Theta waves, enhancing creativity and problem-solving. Theta wave activity can also be enhanced with practices like yoga nidra or hypnosis, which allow access to the subconscious mind.

- **For Delta Wave Balance**: Techniques that promote deep sleep and relaxation, such as progressive relaxation or deep breathing exercises before bed, can help increase Delta wave activity during sleep. Using guided meditation or sound frequencies designed to enhance Delta waves (such as binaural beats) can also be effective.

## 2. Neurofeedback Training

Neurofeedback is a cutting-edge technique used to train individuals to regulate their brainwaves in real time. This method involves the use of EEG devices that monitor brainwave activity while providing feedback to the user. Through this feedback, individuals can learn to increase or decrease specific brainwave frequencies, such as enhancing Alpha waves for relaxation or reducing Beta waves for better focus.

Neurofeedback can be used to treat a variety of conditions caused by brainwave imbalances, including ADHD, anxiety, and insomnia. The goal of neurofeedback training is to help individuals achieve a balanced, healthy brainwave pattern that promotes optimal cognitive performance and emotional regulation.

### 3. Brainwave Entrainment

Brainwave entrainment is a technique that uses rhythmic auditory, visual, or tactile stimuli to synchronize brainwaves to a desired frequency. For example, binaural beats, which involve playing two slightly different frequencies in each ear, can be used to stimulate Alpha, Beta, Theta, or Delta waves. Isochronic tones, a form of sound frequency therapy, can also help promote specific brainwave states.

- **For Relaxation (Alpha/Theta)**: Binaural beats or isochronic tones at 8–12 Hz (Alpha range) can help individuals relax and reduce stress. If creativity or deep meditation is the goal, Theta wave entrainment (4–8 Hz) can be used.

- **For Focus (Beta)**: If concentration or mental alertness is desired, entrainment with Beta waves (12–30 Hz) can improve cognitive performance and problem-solving skills.

- **For Sleep (Delta)**: To promote restful sleep, isochronic tones or binaural beats in the Delta range (0.5–4 Hz) can help stimulate deep sleep and improve restorative processes.

## 4. Lifestyle Adjustments

In addition to meditation, neurofeedback, and brainwave entrainment, lifestyle changes can have a profound impact on brainwave regulation. Practices that support healthy sleep hygiene, stress management, and physical activity can help restore balance between different brainwave states.

- **Sleep Hygiene**: Ensuring adequate and restful sleep is essential for proper Delta wave production. Establishing a consistent sleep routine, limiting screen time before bed, and creating a calming nighttime environment can enhance Delta wave activity during sleep.

- **Physical Exercise**: Regular physical activity, particularly aerobic exercise, can help regulate Beta wave activity, enhancing alertness and focus while reducing stress and anxiety.

- **Nutrition and Hydration**: Consuming a balanced diet rich in Omega-3 fatty acids, B vitamins, and antioxidants can support brain health and promote optimal brainwave functioning. Staying well-hydrated also helps maintain cognitive performance.

## Addressing Conditions Like ADHD, Insomnia, and Anxiety with Brainwave Techniques

Brainwave imbalances are often linked to conditions such as ADHD, insomnia, and anxiety. Fortunately, techniques like neurofeedback, meditation, and brainwave entrainment have shown significant promise in addressing these issues by helping individuals regulate their brainwaves and restore balance.

- **ADHD**: Neurofeedback and mindfulness-based practices are particularly effective in helping individuals with ADHD reduce hyperactivity, improve focus, and increase self-regulation by balancing Beta and Theta wave activity.
- **Insomnia**: Individuals with insomnia often experience imbalances in Delta and Alpha wave activity. Using brainwave entrainment to enhance Delta wave production before sleep or employing relaxation techniques like progressive muscle relaxation can improve sleep quality.
- **Anxiety**: Anxiety is often associated with high Beta wave activity. Meditation, breathing exercises, and neurofeedback can help reduce excessive Beta wave activity, promoting a calm and balanced state of mind.

### Conclusion

Restoring balance to your brainwave activity is crucial for mental and physical health. By identifying brainwave imbalances and using techniques such as meditation, neurofeedback, brainwave entrainment, and lifestyle changes, you can optimize brain function, enhance cognitive performance, and improve emotional well-being. By mastering these techniques, you can regain control over your brainwaves and unlock your full potential.

# Chapter 21: Using Brainwaves to Achieve Flow States

### Understanding Flow States and Their Connection to Brainwave Activity

Flow states, often described as being "in the zone," refer to moments of deep immersion in an activity where a person is fully engaged, focused, and performing at their peak. During flow, time seems to pass unnoticed, and performance tends to be highly efficient and effortless. These moments of optimal experience are not only rewarding but can also lead to significant improvements in creativity, productivity, and overall well-being.

Achieving flow states involves a specific interaction between brainwaves, and the ability to enter these states can be enhanced by consciously regulating brainwave activity. Different types of brainwaves—Alpha, Beta, Delta, and Theta—play crucial roles in creating the ideal mental environment for flow. By understanding how to manipulate brainwaves and align them with the flow state, individuals can significantly enhance their performance in both cognitive and physical domains.

## The Role of Brainwaves in Flow States

Flow is characterized by a high degree of synchronization between different brainwave frequencies. The interaction of Alpha, Beta, Delta, and Theta waves creates an optimal state of cognitive and emotional balance. Understanding how each of these brainwaves contributes to flow is key to mastering the state.

### Alpha Waves (8–12 Hz): The Gateway to Relaxed Focus

Alpha waves are typically associated with relaxation and calmness. However, they also play an essential role in creating the mental clarity required for flow. When Alpha waves dominate, the brain is in a state of relaxed focus, which allows individuals to remain alert without feeling overly stressed or distracted. This relaxed state is crucial for entering flow because it helps to reduce mental clutter while maintaining enough focus to engage in the task at hand.

To reach a flow state, it is essential to increase Alpha wave activity without descending into deep relaxation or drowsiness. Techniques such as mindfulness meditation, deep breathing, and visualization exercises can increase Alpha wave production and create the right mental conditions for flow.

## Beta Waves (12–30 Hz): The Drive for Focused Attention

Beta waves are linked to active concentration, problem-solving, and engagement in tasks that require attention. However, too much Beta activity can lead to anxiety and mental fatigue, potentially hindering the ability to enter flow. For flow to occur, Beta waves need to be elevated just enough to foster intense focus without pushing the mind into overdrive.

Beta waves are most beneficial in flow when they interact harmoniously with Alpha waves. This balance allows for optimal focus while maintaining mental clarity and reducing unnecessary stress. Brainwave entrainment, cognitive training, and deliberate practice in focused tasks can help increase Beta wave activity within the range that promotes peak performance.

### Theta Waves (4–8 Hz): The Key to Creativity and Deep Immersion

Theta waves, which are associated with deep relaxation, creativity, and access to the subconscious, play a significant role in flow. Theta activity is commonly linked to states of deep meditation, where the mind becomes receptive to new ideas and innovative solutions. In flow, Theta waves allow the mind to tap into intuitive thought processes and creative insights, often leading to novel solutions or breakthroughs in problem-solving.

The challenge with Theta waves in flow states is that while they promote creativity and deep immersion, they can also lead to distraction or disengagement if they become too dominant. Balancing Theta waves with Alpha and Beta waves helps maintain the creativity and relaxed focus necessary for flow without losing the ability to concentrate on the task at hand.

### Delta Waves (0.5-4 Hz): Restorative and Healing

Although Delta waves are most associated with deep sleep and restorative processes, they can also play a role in achieving flow. Delta waves help to facilitate the body's healing processes and contribute to overall mental and physical well-being. In certain instances, a person in flow may experience subtle Delta wave activity, particularly if they are engaged in highly immersive and physically demanding tasks.

For instance, athletes in peak performance might enter a state where their body and mind are in perfect harmony, supported by the slow, restorative Delta waves that foster deep relaxation and recovery. However, for most flow activities, Delta wave activity should be minimal to avoid mental sluggishness.

## Techniques to Induce Flow States Using Brainwaves

Achieving flow requires the ability to regulate brainwave activity consciously and use different techniques to promote optimal states of mind. There are several ways to induce and sustain flow states using brainwaves.

### 1. Meditation and Mindfulness

Meditation is a powerful tool for regulating brainwave activity and achieving flow. Mindfulness meditation, in particular, helps to increase Alpha wave activity while calming Beta and Theta waves. By practicing mindfulness, individuals can quiet the mind, reduce distractions, and enter a relaxed yet focused state—ideal for achieving flow.

For creative tasks or problem-solving, visualization techniques can be used in meditation to enhance Theta wave activity, allowing individuals to tap into subconscious insights and creative ideas. Combining meditation with deep breathing or progressive muscle relaxation also helps to control Beta wave activity and maintain an optimal balance between alertness and relaxation.

## 2. Neurofeedback Training

Neurofeedback is a brain training technique that uses real-time feedback from brainwave measurements (usually via EEG) to help individuals learn to regulate their brainwave activity. Through neurofeedback, individuals can train their brains to enhance Alpha and Beta waves while modulating Theta waves to achieve a balanced state.

Neurofeedback training can be particularly effective for those seeking to enter flow states on command. By training the brain to maintain the optimal brainwave frequencies for flow, neurofeedback helps individuals develop the ability to shift into peak performance mode more easily.

## 3. Brainwave Entrainment

Brainwave entrainment uses rhythmic auditory, visual, or tactile stimuli to influence brainwave patterns. Binaural beats, isochronic tones, and light frequencies are common forms of brainwave entrainment. These tools can be used to guide the brain into the desired frequency range for flow.

- **Binaural beats**: Two slightly different sound frequencies played in each ear stimulate the brain to synchronize its activity to the difference in frequencies. For flow, binaural beats in the Alpha (8-12 Hz) or Theta (4-8 Hz) range are most effective.
- **Isochronic tones**: These are single tones that pulse at specific intervals, which can induce brainwave entrainment. Isochronic tones for flow typically range from Alpha to Theta frequencies, depending on the task.

Using these auditory tools regularly can help an individual condition their brain to enter and sustain flow states more easily.

## 4. Physical Exercise and Body Awareness

Physical exercise, particularly aerobic activity, is known to increase Beta wave activity and enhance overall brain function. Exercise can serve as a form of flow induction by fostering focus, attention, and mental clarity. The endorphins released during physical activity can further enhance mood and reduce stress, making it easier to enter flow.

Yoga and other mindful movement practices are also powerful tools for inducing flow states. The combination of deep breathing, focused attention, and body awareness helps regulate brainwave patterns, bringing them into an optimal state for flow.

## 5. Optimal Challenge and Skill Balance

Flow is most likely to occur when the challenge of an activity matches an individual's skill level. If the task is too easy, the person may become bored and disengaged, while if the task is too difficult, anxiety may interfere with performance. By adjusting the difficulty level of tasks to match one's abilities, individuals can increase their chances of experiencing flow.

This balance between challenge and skill is critical to entering flow and maintaining it. By gradually increasing the challenge of a task as skills improve, individuals can keep their brain in the optimal state for flow, where they experience both deep focus and enjoyment.

## How Athletes and Performers Achieve Peak Concentration and Performance

Flow states are common among athletes, musicians, and other performers. In these high-performance environments, the ability to regulate brainwave activity is crucial. Athletes in flow experience heightened concentration, swift decision-making, and physical grace. Brainwave patterns in flow states often show a balance of Alpha and Beta waves, allowing for both relaxation and intense focus.

For athletes, mental preparation techniques such as visualization, relaxation exercises, and focused breathing can induce Alpha and Theta waves, making it easier to reach a flow state during physical activity. Similarly, musicians and performers use mindfulness and practice to enter the zone where their skills align seamlessly with the task at hand.

## Conclusion

Achieving flow states through brainwave regulation is a powerful tool for enhancing both cognitive and physical performance. By understanding the role of Alpha, Beta, Theta, and Delta waves in flow, individuals can use techniques like meditation, neurofeedback, brainwave entrainment, and physical exercise to optimize their mental state. Whether for work, sports, or creative endeavors, mastering the ability to enter flow can lead to peak performance, increased productivity, and a more fulfilling life.

# Chapter 22: The Role of Meditation in Brainwave Mastery

## The Importance of Meditation in Controlling Brainwaves

Meditation has been practiced for thousands of years for its profound ability to enhance mental clarity, emotional stability, and overall well-being. Over time, research has uncovered the remarkable impact meditation can have on brainwave activity, showing how it can help individuals master their neural oscillations for optimal performance. The practice of meditation, whether through mindfulness, focused attention, or deep relaxation, offers a unique way to regulate brainwave frequencies intentionally, cultivating the states most conducive to achieving mental, emotional, and physical balance.

When you meditate, your brain is encouraged to shift from higher-frequency Beta waves (associated with active thinking and alertness) to lower-frequency Alpha, Theta, and even Delta waves, depending on the depth and type of meditation. This shift in brainwave activity helps to induce a state of focused calm, reduce stress, promote healing, and unlock creativity—allowing you to achieve a more balanced and connected state of mind.

## How Different Meditation Techniques Affect Brainwave States

There are numerous forms of meditation, each with its own distinct approach and corresponding effects on brainwave activity. Here are some of the most widely practiced techniques and how they influence brainwaves:

### 1. Mindfulness Meditation (Alpha and Theta Waves)

Mindfulness meditation, perhaps the most popular meditation technique in modern wellness practices, encourages a state of awareness in the present moment. This practice has been shown to increase Alpha wave activity (8-12 Hz), which is associated with a relaxed, yet alert, state of mind. It also enhances Theta waves (4-8 Hz), which are linked to deeper states of relaxation and access to the subconscious mind.

Mindfulness meditation helps reduce stress and anxiety by calming the constant chatter of the mind, allowing for greater clarity and focus. It's particularly effective in boosting emotional resilience, improving cognitive function, and fostering self-awareness. As practitioners become more skilled, Theta waves become more prominent, allowing for a deeper sense of calm, creativity, and innovation.

## 2. Concentrative Meditation (Beta Waves)

In contrast to mindfulness meditation, concentrative meditation involves focusing on a single object or thought, such as the breath, a mantra, or a visual image. This type of meditation tends to activate Beta waves (12-30 Hz), which are associated with alertness and concentration. Though higher-frequency Beta waves are often linked with stress and overthinking, concentrative meditation can help fine-tune these waves to a productive level, facilitating enhanced focus and attention.

Practicing this type of meditation enables individuals to sharpen their concentration, improve mental clarity, and enhance problem-solving skills. It's especially useful for tasks that require sustained attention and cognitive precision.

### 3. Loving Kindness Meditation (Alpha Waves)

Loving Kindness Meditation, or *Metta Bhavana*, is a practice centered around cultivating compassion, love, and kindness towards oneself and others. This practice has been shown to activate Alpha waves (8-12 Hz), promoting a state of calm and emotional balance. As you meditate on positive feelings of compassion, Alpha waves help induce a relaxed, open-hearted state that encourages emotional healing and well-being.

This technique not only fosters positive emotional states but also reduces stress and anxiety, increases feelings of empathy, and improves interpersonal relationships. By balancing emotional responses and promoting a state of inner peace, Loving Kindness Meditation helps individuals regulate their emotional and physical well-being.

### 4. Transcendental Meditation (Alpha and Theta Waves)

Transcendental Meditation (TM) involves repeating a mantra to quiet the mind and reach a transcendent state of consciousness. Research on TM has shown that it induces both Alpha and Theta waves, which are associated with deep relaxation and heightened awareness. This meditation technique is particularly effective at reducing stress and promoting overall mental and physical health.

Practitioners of TM report increased focus, clarity, and creativity, as well as improvements in emotional regulation. By accessing deeper levels of brainwave activity, TM facilitates a profound sense of stillness and inner peace, helping individuals reduce mental clutter and achieve a state of deep concentration and relaxation.

### 5. Yoga and Breathwork (Alpha and Theta Waves)

Yoga, particularly when combined with mindful breathing techniques (Pranayama), can stimulate Alpha and Theta waves. The controlled movements, deep stretches, and focus on breath not only benefit the body but also help synchronize brainwave activity, bringing it into balance. Certain yoga practices, like restorative yoga and yoga nidra, induce a state of deep relaxation, promoting the release of tension and fostering a sense of inner calm.

Breathing exercises such as slow, deep abdominal breathing can also help lower the frequency of Beta waves, reducing anxiety and stress. Breathwork techniques, especially when combined with yoga postures, regulate the autonomic nervous system and create a harmonious balance between body and mind, making it easier to enter a meditative state conducive to deep relaxation and heightened awareness.

## Creating a Daily Meditation Routine for Brainwave Optimization

Developing a consistent meditation practice is one of the most effective ways to optimize brainwave activity for personal growth and well-being. By regularly practicing meditation, you can train your brain to enter specific brainwave states on demand, improving your ability to manage stress, enhance focus, and foster emotional resilience. Here's how to create a daily meditation routine:

### 1. Set Your Intention

Before you begin meditating, set an intention for your practice. Whether it's to reduce stress, improve focus, or cultivate creativity, having a clear purpose can help guide your practice and keep you focused. The brain responds well to intentionality, which can enhance the benefits of meditation.

### 2. Find a Comfortable Space

Choose a quiet and comfortable space where you won't be disturbed. Make sure the environment is free from distractions, and set up a space where you feel relaxed and grounded. This could be a corner in your home, a park, or even a dedicated meditation space.

### 3. Focus on Your Breath or Mantra

Start by focusing on your breath or silently repeating a mantra. Focusing on the breath encourages Alpha and Theta wave activity, while mantras can help direct concentration and foster calmness. If your mind wanders, gently bring it back to your chosen point of focus.

### 4. Experiment with Different Techniques

Try different types of meditation to see what works best for you. You might find that mindfulness meditation helps you relax, while concentrative meditation boosts your focus. As you become more experienced, you can tailor your practice to suit your specific needs.

### 5. Gradually Increase Duration

Start with just 5-10 minutes of meditation each day, and gradually increase the duration as you become more comfortable with the practice. Over time, you'll be able to enter meditative states more easily and quickly, enhancing the benefits of your brainwave training.

## Conclusion

Meditation is a powerful tool for mastering brainwave activity, offering a pathway to improved mental, emotional, and physical well-being. By practicing different meditation techniques and cultivating a consistent routine, you can fine-tune your brainwaves to achieve the optimal state for relaxation, focus, creativity, and peak performance. Whether you're seeking to reduce stress, boost productivity, or enhance creativity, meditation provides the means to unlock your brain's full potential and improve your overall quality of life.

# Chapter 23: Advanced Techniques for Brainwave Control

### Introduction: Going Beyond the Basics

As we've explored in earlier chapters, brainwave activity is a powerful tool for personal growth, cognitive enhancement, and emotional well-being. While basic techniques such as meditation, mindfulness, and neurofeedback can help regulate brainwaves, advanced methods allow individuals to take control over their brainwaves in more specific and nuanced ways. These techniques enable deeper mastery of neural oscillations and pave the way for optimizing brain function, achieving peak performance, and fostering creativity.

In this chapter, we will dive into advanced techniques for brainwave control, including the use of brainwave entrainment, neurofeedback training, and cutting-edge technologies. These methods provide not only a deeper understanding of brainwave behavior but also the tools to intentionally influence brain activity in real-time, which can be particularly beneficial for achieving specific states such as creativity, focus, relaxation, or deep restorative sleep.

## 1. Advanced Brainwave Entrainment Techniques

Brainwave entrainment is the process of using external stimuli (sound, light, or electromagnetic fields) to synchronize brainwave frequencies to a desired state. While basic entrainment techniques like binaural beats and isochronic tones have been widely discussed, advanced entrainment methods can fine-tune the brain's oscillatory patterns even further, enabling greater specificity in the brainwave state you wish to achieve.

## Isochronic Tones and Pulsed Light

Isochronic tones are single tones that pulse at a specific frequency, used to encourage the brain to match its own oscillations to the frequency of the tone. Unlike binaural beats, which require the use of headphones to create the perception of a beat, isochronic tones can be heard through regular speakers and are more powerful in entraining brainwaves because of their clear and direct pulses.

Pulsed light is another effective method used for advanced entrainment. Light pulses, particularly those in the alpha and theta ranges, can stimulate the brain to produce corresponding waves. These light frequencies can be delivered through devices like light goggles or specialized lamps. The visual stimuli combined with specific frequencies help deepen meditation, promote relaxation, and enhance creativity.

### Frequency Following Response (FFR)

The Frequency Following Response (FFR) is a phenomenon where the brain adjusts its own frequency to match the frequency of external stimuli. Advanced brainwave entrainment techniques use this natural response to guide brain activity into specific states, such as deep meditation (theta waves) or heightened alertness (beta waves). Researchers are continually refining the use of FFR through more precise and tailored stimuli that target individual needs, whether for relaxation or cognitive performance.

## 2. Neurofeedback Training: Precision Brainwave Regulation

Neurofeedback is a powerful technique that allows individuals to learn how to regulate their brainwave activity by providing real-time feedback on their brain's electrical activity, usually through an EEG. This process involves training the brain to increase or decrease certain types of brainwaves, promoting a more balanced and optimal state of mind.

### EEG Neurofeedback

EEG neurofeedback is one of the most popular neurofeedback techniques, and it works by measuring brainwave activity through sensors placed on the scalp. This data is then fed back to the user through visual, auditory, or tactile cues, allowing them to see or hear when they are entering the desired brainwave state. Over time, with repeated training, the brain learns to produce the target frequency more readily.

EEG neurofeedback is particularly effective in treating conditions like ADHD, anxiety, depression, and insomnia, where imbalances in specific brainwave frequencies may be present. Advanced neurofeedback programs can train individuals to enhance alpha waves for relaxation, beta waves for concentration, or theta waves for creativity and deep meditation.

## Real-time Biofeedback: A Holistic Approach

In addition to EEG neurofeedback, there are newer techniques involving other types of biofeedback, such as heart rate variability (HRV) and respiration monitoring. These methods allow for a more holistic approach to brainwave regulation by integrating the body's physiological signals with the brain's electrical activity. For example, HRV biofeedback can help synchronize heart rate with breath and brainwave activity, promoting calmness and reducing stress.

Real-time biofeedback has been shown to enhance cognitive performance, lower stress levels, and increase emotional resilience. With these methods, individuals are empowered to regulate not just their brainwaves, but their entire physiological state.

## 3. Advanced Wearables and Brain–Computer Interfaces (BCIs)

The field of brainwave mastery has evolved significantly with the advent of wearable technology and brain-computer interfaces (BCIs). These cutting-edge devices provide a hands-on approach to influencing and optimizing brain activity by using direct brainwave monitoring and real-time feedback. BCIs are particularly important in professional, clinical, and research settings where precision brainwave manipulation is required.

## BCI Technology for Brainwave Mastery

Brain-computer interfaces (BCIs) allow for direct communication between the brain and an external device, such as a computer, prosthetic limb, or virtual environment. While BCIs are often associated with medical uses (such as helping individuals with disabilities), they are also increasingly used in research and development for enhancing cognitive performance. By detecting and interpreting specific brainwave patterns, BCIs can help individuals train their brains to optimize mental states, from deep concentration to creative flow.

In the context of brainwave mastery, BCIs can be used for a wide range of applications, including cognitive enhancement, stress management, and meditation. For example, a BCI device can be used to measure theta wave activity in real time and provide immediate feedback, allowing the user to learn how to enter deeper meditative states at will. This direct interaction with brainwaves provides a powerful method for fine-tuning brain activity for personal growth.

## Advanced Wearables for Cognitive Enhancement

Wearables, such as EEG headsets, are becoming more sophisticated, allowing individuals to track their brainwave activity throughout the day. These devices can provide ongoing feedback, encouraging users to regulate their brainwaves on the fly. For instance, if a wearable device detects an excess of beta waves (indicating stress or overthinking), it can prompt the wearer to take a break or engage in a relaxation technique to bring alpha waves back into balance.

Some advanced wearables even offer integrated training programs that guide users through exercises designed to enhance specific cognitive functions, like improving focus, reducing stress, or enhancing creativity. These wearable technologies offer users a powerful, personalized tool to achieve and maintain optimal brainwave states.

## 4. Neurostimulation and Transcranial Magnetic Stimulation (TMS)

Neurostimulation techniques like Transcranial Direct Current Stimulation (tDCS) and Transcranial Magnetic Stimulation (TMS) are advanced methods for influencing brainwave activity. These techniques use external electrical or magnetic fields to directly stimulate brain regions, influencing neural oscillations.

## Transcranial Direct Current Stimulation (tDCS)

tDCS involves applying a low electrical current to the scalp using electrodes. This current is designed to modulate the activity of specific brain regions, either enhancing or inhibiting their activity. This technique has been shown to increase the performance of cognitive tasks by altering brainwave activity in targeted regions of the brain. tDCS can be used to improve learning, enhance memory, and facilitate focus.

## Transcranial Magnetic Stimulation (TMS)

TMS uses magnetic pulses to stimulate specific areas of the brain. This non-invasive technique is particularly useful for altering brainwave frequencies in specific regions and has been applied in clinical settings to treat depression and other mental health conditions. TMS is also being explored for cognitive enhancement, as it has the potential to induce targeted brainwave states that can lead to improved performance in tasks requiring concentration or creativity.

## Conclusion: Mastering Brainwave Control

Advanced techniques for brainwave control provide powerful tools for unlocking the full potential of the human brain. Whether through advanced brainwave entrainment, neurofeedback, wearable technology, or neurostimulation, these methods allow for a deeper level of mastery over brainwave activity. By learning to regulate brainwaves with precision, individuals can improve cognitive function, emotional regulation, and overall mental health. The combination of these cutting-edge techniques represents the frontier of brainwave mastery, offering unprecedented opportunities for growth, performance, and well-being.

# Chapter 24: The Future of Brainwave Mastery

### Introduction: The Frontier of Brainwave Research

Over the course of this book, we have explored the foundational concepts of brainwave activity, its impact on cognition, emotional regulation, and physical performance, as well as techniques for mastering neural oscillations. We have looked at practical methods such as meditation, mindfulness, neurofeedback, and brainwave entrainment, all of which allow individuals to regulate their brainwaves for optimal well-being. However, the field of brainwave mastery is rapidly advancing. As technology continues to evolve, new frontiers in brainwave research are emerging, offering exciting possibilities for more precise, efficient, and personalized brainwave optimization.

In this chapter, we will explore the cutting-edge technologies, trends, and potential developments that will shape the future of brainwave mastery. We will look at the role of artificial intelligence (AI) and brain-computer interfaces (BCIs) in brainwave regulation, the promise of wearable devices, and the ethical considerations that come with controlling brainwaves on such an intricate level. The future of brainwave mastery is not just about improving cognitive performance—it has the potential to revolutionize mental health treatment, self-development, and human augmentation.

## 1. The Rise of Artificial Intelligence and Machine Learning

Artificial intelligence (AI) is already making a significant impact on brainwave research. By combining AI with brainwave data, we can now analyze vast amounts of information and identify patterns that were previously undetectable by the human eye. AI has the potential to offer highly personalized brainwave training by using machine learning algorithms to adapt the training process to each individual's unique brainwave patterns.

## AI-Driven Brainwave Optimization

AI can be used to create dynamic, real-time feedback systems that adjust brainwave training protocols based on the individual's response. For example, AI could continuously analyze EEG data to determine the most effective frequency for promoting relaxation, focus, or creativity, adjusting stimuli (e.g., sound, light, or vibration) in real-time to optimize the brain's oscillatory state.

This level of sophistication in brainwave entrainment and neurofeedback could allow for more precise control over brainwave activity, enabling individuals to access and maintain desired brainwave states with greater ease and consistency. Additionally, AI could help to develop advanced training programs that focus not just on isolated states like alpha or theta waves, but on the dynamic interplay between different brainwave frequencies for improved overall cognitive performance.

## Neuroadaptive Technologies

The combination of AI and neurofeedback is paving the way for neuroadaptive technologies. These systems use AI to adapt to the brain's needs in real-time, fine-tuning brainwave states as they fluctuate. Imagine a system that not only helps you enter a deep meditative state (e.g., via theta waves) but also guides you toward greater mental clarity and focus as your brain transitions to a more alert state (e.g., beta waves) as the need arises.

This neuroadaptive feedback could be integrated into wearable technologies or specialized apps, providing a seamless user experience that helps individuals enhance their mental states throughout the day, increasing focus and productivity while managing stress.

## 2. The Role of Brain–Computer Interfaces (BCIs) in Brainwave Mastery

Brain-Computer Interfaces (BCIs) have garnered significant attention in recent years for their potential to bridge the gap between the brain and external devices, allowing users to control technology with their thoughts. BCIs can be used to detect brainwave activity and translate it into commands that control devices such as computers, prosthetics, and even virtual reality (VR) environments.

## BCIs for Brainwave Regulation

In the context of brainwave mastery, BCIs are poised to become a central tool in the real-time regulation of brain activity. By directly interfacing with the brain, BCIs can provide detailed insights into brainwave patterns and deliver feedback that encourages users to enhance or diminish specific frequencies. This could be used for therapeutic purposes (e.g., managing depression, anxiety, or ADHD) as well as cognitive enhancement (e.g., increasing focus, memory retention, or creativity).

BCIs could also allow for more precise control of brainwave states, even allowing individuals to "program" their brains for specific tasks. For instance, a BCI could help you transition between alpha and beta waves, optimizing your state of mind for both relaxation and mental alertness, depending on your needs.

## Brainwave-Specific BCIs

As BCI technology advances, it may evolve into systems designed to target specific brainwave frequencies more effectively. For example, a BCI system could be optimized to enhance alpha wave activity for stress relief and relaxation, while another could focus on beta waves for improving mental performance. Over time, BCIs could become an essential tool in personalizing brainwave mastery, with users able to select the exact brainwave patterns they want to cultivate in real-time.

### 3. Wearable Devices and Biofeedback Systems

The wearable technology industry is rapidly advancing, and new devices are being created that allow individuals to monitor and control their brainwaves in real-time. These wearables are designed to be both discreet and effective, allowing for continuous monitoring of brain activity without the need for cumbersome equipment.

### EEG Headsets and Brainwave Monitoring

EEG headsets, which were once primarily used in clinical settings, are now becoming more accessible to consumers. These devices are designed to detect brainwave activity using sensors placed on the scalp, providing real-time data about brain state. Advanced EEG wearables can help individuals track their progress, measure the effects of different training techniques, and optimize their brainwave patterns for specific goals.

With real-time feedback, users can quickly adjust their behavior, such as using breathing techniques or mindfulness practices, to bring their brainwaves into balance. Some wearables even provide visual or auditory cues that help users maintain or transition between specific brainwave states, encouraging a seamless integration of brainwave optimization into daily life.

## Smart Headbands and VR for Brainwave Control

In addition to EEG headsets, smart headbands and virtual reality (VR) systems are increasingly being used to enhance brainwave control. These devices combine brainwave monitoring with VR environments or auditory feedback, offering users a more immersive experience in brainwave mastery. For example, VR platforms that incorporate real-time brainwave data could guide individuals through scenarios where they are encouraged to achieve certain brainwave states for relaxation, focus, or creativity.

## 4. Ethical Considerations of Brainwave Control

With the rapid advancements in brainwave mastery technologies, ethical considerations become crucial. As we gain the ability to control our brainwaves with increasing precision, questions arise about the potential for misuse or over-reliance on these technologies.

## The Ethical Use of Neurofeedback and BCIs

While BCIs and neurofeedback devices have the potential to greatly enhance cognitive performance and mental health, they also raise concerns about privacy, autonomy, and consent. How can we ensure that individuals are not manipulated or coerced into modifying their brainwaves for external purposes? Additionally, there are concerns about the potential for "brain hacking" or using these technologies for unethical purposes, such as controlling thoughts, behaviors, or emotions.

## Balancing Brainwave Mastery with Personal Autonomy

As with any powerful technology, balancing the potential benefits of brainwave control with the preservation of personal autonomy is critical. It is essential to ensure that individuals maintain control over their own cognitive and emotional states while benefiting from these technologies. Ensuring that brainwave mastery remains a tool for empowerment, rather than a mechanism for manipulation, will be a key challenge in the coming years.

## Conclusion: Embracing the Future of Brainwave Mastery

The future of brainwave mastery is filled with exciting possibilities. As AI, BCIs, wearables, and other advanced technologies continue to evolve, they will provide individuals with unprecedented levels of control over their cognitive and emotional states. However, with these advancements come important ethical considerations that must be addressed to ensure these technologies are used responsibly.

As you embark on your journey of brainwave mastery, remember that these emerging technologies offer an exciting opportunity to enhance your mental, emotional, and physical well-being. By understanding the future of brainwave research, you can make informed decisions about how to incorporate these tools into your life while maintaining a sense of personal autonomy and ethical responsibility. The future of brainwave mastery is not just about achieving optimal brain function—it's about harnessing the power of your mind to unlock your full potential.

# Chapter 25: Conclusion: Unlocking Your Full Potential Through Brainwave Mastery

### The Journey to Mastering Your Mind

Throughout this book, we have explored the fascinating world of neural oscillation and brainwaves. We have examined how these electrical patterns within the brain—Alpha, Beta, Delta, and Theta waves—play a vital role in shaping our cognition, behavior, emotional responses, and overall well-being. From understanding the basics of neural oscillations to employing sophisticated tools like neurofeedback and brainwave entrainment, we have covered a range of strategies to harness the power of your brain's natural rhythms.

Mastering brainwaves is not simply about improving performance in one area of life. It is about achieving a higher state of balance, awareness, and control that permeates every aspect of your mental, emotional, and physical health. It is about unlocking your full potential, gaining deeper self-awareness, and using the power of your mind to create positive transformations in your life.

In this final chapter, we will summarize the key concepts and techniques you have learned, explore how to implement them into your daily life, and offer guidance on creating a personal plan for ongoing brainwave mastery. The journey to mastering your brainwaves is ongoing—one of continuous growth, self-discovery, and mastery.

## 1. Recap of Key Concepts and Techniques

As we've seen, brainwaves are the fundamental building blocks of mental and emotional states. Their different frequencies—Alpha, Beta, Delta, and Theta—are not isolated from one another but interact, influencing one another dynamically. By learning to regulate and optimize these waves, you can:

- **Enhance Cognitive Function**: With techniques like brainwave entrainment, meditation, and mindfulness, you can improve memory, focus, and learning. Alpha and Beta waves, for example, help you manage stress while enhancing cognitive performance, while Theta waves can deepen creativity and problem-solving.

- **Promote Mental Health**: Brainwave regulation is a powerful tool for managing mental health. Through practices such as mindfulness, neurofeedback, and meditation, you can restore balance between your brainwave states and mitigate symptoms of anxiety, depression, and other mental health disorders.

- **Optimize Physical Performance**: Brainwaves also impact physical performance. By using Alpha waves to relax, Beta waves to stay alert, and Theta waves to access deeper creative or problem-solving states, athletes, performers, and professionals can boost their physical and mental performance.

- **Leverage Sleep and Healing**: Delta waves play a critical role in deep sleep and restorative processes. By optimizing Delta wave activity, you can promote better sleep, aid tissue repair, and enhance your overall health and healing.

## 2. Implementing Brainwave Mastery in Daily Life

Now that we've covered the science and tools, how can you apply these insights into your daily life? Mastering your brainwaves doesn't require radical changes but can be woven into your routine in ways that significantly enhance your well-being.

### Creating a Brainwave Training Routine

Consider incorporating brainwave optimization techniques into your daily activities. For example:

- **Morning Focus**: Start your day by engaging in a short meditation or mindfulness session to enhance Alpha or Beta wave activity, promoting a calm, focused mind.
- **Midday Productivity Boost**: Use binaural beats or isochronic tones to induce Beta waves during work or study, enhancing productivity and cognitive performance.
- **Evening Relaxation**: Wind down at the end of your day with Theta wave meditation, preparing your mind for restful sleep. You can also incorporate Delta-focused techniques to improve sleep quality.

## Using Neurofeedback Devices

Neurofeedback is an excellent tool to track your brainwave activity and improve your ability to regulate it. Using devices like EEG headsets, you can receive real-time feedback about your brainwave patterns, helping you learn to consciously shift between different states. Over time, you can use neurofeedback to train your brain to enter optimal states more quickly and effectively.

## Mindful Movement and Physical Training

Physical exercise, yoga, and other mindful movement practices can also be used to harmonize brainwave activity. Certain movements or breathwork can activate different brainwave states, providing both physical and mental benefits. For example, deep breathing exercises can help you enter an Alpha state, reducing stress and promoting relaxation.

### 3. Creating Your Personal Brainwave Mastery Plan

Brainwave mastery is a lifelong journey. The first step to continuing this path is to create a personalized plan that aligns with your specific goals. Here are some steps to guide you:

### Assess Your Current Brainwave Activity

Start by reflecting on your current mental and emotional states. Are you often stressed and anxious? Do you struggle with focus or creativity? Identify areas where brainwave optimization could have the most impact.

### Set Your Goals

What do you hope to achieve with brainwave mastery? Whether it's improving concentration, enhancing creativity, or improving sleep quality, setting clear goals will guide your brainwave training efforts. Be sure to define both short-term and long-term objectives.

### Choose Your Techniques

Based on your goals, select the brainwave training techniques that will best help you. This could include:

- **Meditation** (to enhance Alpha or Theta activity)
- **Neurofeedback** (to train specific brainwave patterns)
- **Brainwave entrainment** (using sound or light frequencies to guide brainwave activity)
- **Cognitive exercises** (to strengthen Beta waves for focus or problem-solving)

## Track Your Progress

Keep track of your progress through journaling or using brainwave measurement tools. Note any changes in mood, productivity, or overall well-being as you engage in brainwave optimization practices. Over time, you will notice patterns in your brainwave states and how they correlate to different mental and emotional states.

## Stay Consistent and Adjust

As with any form of mastery, consistency is key. Integrate your brainwave training into your daily routine and stick with it. However, be open to adjusting your methods as you learn more about what works best for you. With time, you will refine your practices to achieve more profound results.

## 4. Final Thoughts: Unlocking Your Full Potential

Mastering your brainwaves is a powerful tool for personal growth, but it is only one part of a larger journey toward self-optimization. As you continue exploring the possibilities of brainwave mastery, remember that it is about more than just improving your cognitive abilities. It is about achieving a balanced, peaceful, and empowered life where you can tap into your fullest potential.

The tools and techniques you have learned are just the beginning. As the field of brainwave research continues to evolve, new innovations will emerge, providing even more opportunities for growth and development. Stay curious, keep experimenting, and embrace the ongoing process of mastering your mind.

The key to unlocking your full potential is within you—and it all starts with mastering the frequencies of your brain.